D1267296

Exploring Advanced Euclidean Geometry with GeoGebra

Library of Congress Catalog Card Number 2013938569

Print edition ISBN 978-0-88385-784-7

Electronic edition ISBN 978-1-61444-111-3

Printed in the United States of America

Current Printing (last digit):
10 9 8 7 6 5 4 3 2

Exploring Advanced Euclidean Geometry with GeoGebra

Gerard A. Venema
Calvin College

Published and Distributed by
The Mathematical Association of America

CLASSROOM RESOURCE MATERIALS

Classroom Resource Materials is intended to provide supplementary classroom material for students—laboratory exercises, projects, historical information, textbooks with unusual approaches for presenting mathematical ideas, career information, etc.

101 Careers in Mathematics, 2nd edition edited by Andrew Sterrett

Archimedes: What Did He Do Besides Cry Eureka?, Sherman Stein

Calculus: An Active Approach with Projects, Stephen Hilbert, Diane Driscoll Schwartz, Stan Seltzer, John Maceli, and Eric Robinson

Calculus Mysteries and Thrillers, R. Grant Woods

Conjecture and Proof, Miklós Laczkovich

Counterexamples in Calculus, Sergiy Klymchuk

Creative Mathematics, H. S. Wall

Environmental Mathematics in the Classroom, edited by B. A. Fusaro and P. C. Kenschaft

Excursions in Classical Analysis: Pathways to Advanced Problem Solving and Undergraduate Research, by Hongwei Chen

Explorations in Complex Analysis, Michael A. Brilleslyper, Michael J. Dorff, Jane M. McDougall, James S. Rolf, Lisbeth E. Schaubroeck, Richard L. Stankewitz, and Kenneth Stephenson

Exploratory Examples for Real Analysis, Joanne E. Snow and Kirk E. Weller

Exploring Advanced Euclidean Geometry with GeoGebra, Gerard A. Venema

Geometry From Africa: Mathematical and Educational Explorations, Paulus Gerdes

Historical Modules for the Teaching and Learning of Mathematics (CD), edited by Victor Katz and Karen Dee Michalowicz

Identification Numbers and Check Digit Schemes, Joseph Kirtland

Interdisciplinary Lively Application Projects, edited by Chris Arney

Inverse Problems: Activities for Undergraduates, Charles W. Groetsch

Keeping it R.E.A.L.: Research Experiences for All Learners, Carla D. Martin and Anthony Tongen

Laboratory Experiences in Group Theory, Ellen Maycock Parker

Learn from the Masters, Frank Swetz, John Fauvel, Otto Bekken, Bengt Johansson, and Victor Katz

Math Made Visual: Creating Images for Understanding Mathematics, Claudi Alsina and Roger B. Nelsen

Mathematics Galore!: The First Five Years of the St. Marks Institute of Mathematics, James Tanton

Methods for Euclidean Geometry, Owen Byer, Felix Lazebnik, and Deirdre L. Smeltzer

Ordinary Differential Equations: A Brief Eclectic Tour, David A. Sánchez

Oval Track and Other Permutation Puzzles, John O. Kiltinen

Paradoxes and Sophisms in Calculus, Sergiy Klymchuk and Susan Staples

A Primer of Abstract Mathematics, Robert B. Ash

Proofs Without Words, Roger B. Nelsen

Proofs Without Words II, Roger B. Nelsen

Rediscovering Mathematics: You Do the Math, Shai Simonson

She Does Math!, edited by Marla Parker

Solve This: Math Activities for Students and Clubs, James S. Tanton

Student Manual for Mathematics for Business Decisions Part 1: Probability and Simulation, David Williamson, Marilou Mendel, Julie Tarr, and Deborah Yoklic

Student Manual for Mathematics for Business Decisions Part 2: Calculus and Optimization, David Williamson, Marilou Mendel, Julie Tarr, and Deborah Yoklic

Teaching Statistics Using Baseball, Jim Albert

Visual Group Theory, Nathan C. Carter

Which Numbers are Real?, Michael Henle

Writing Projects for Mathematics Courses: Crushed Clowns, Cars, and Coffee to Go, Annalisa Crannell, Gavin LaRose, Thomas Ratliff, and Elyn Rykken

MAA Service Center
P.O. Box 91112
Washington, DC 20090-1112
1-800-331-1MAA FAX: 1-301-206-9789

Preface

This book provides an inquiry-based introduction to advanced Euclidean geometry. It can be used either as a computer laboratory manual to supplement a course in the foundations of geometry or as a stand-alone introduction to advanced topics in Euclidean geometry. The geometric content is substantially the same as that of the first half of the classic text *Geometry Revisited* by Coxeter and Greitzer [3]; the organization and method of study, however, are quite different. The book utilizes dynamic geometry software, specifically GeoGebra, to explore the statements and proofs of many of the most interesting theorems in advanced Euclidean geometry. The text consists almost entirely of exercises that guide students as they discover the mathematics and then come to understand it for themselves.

Geometric content

The geometry studied in this book is Euclidean geometry. Euclidean geometry is named for Euclid of Alexandria, who lived from approximately 325 BC until about 265 BC. The ancient Greeks developed geometry to a remarkably advanced level and Euclid did his work during the later stages of that development. He wrote a series of books, called the *Elements*, that organize and summarize the geometry of ancient Greece. Euclid's *Elements* became by far the best known geometry text in history and Euclid's name is universally associated with geometry as a result.

Roughly speaking, *elementary* Euclidean geometry is the geometry that is contained in Euclid's writings. Most readers will already be familiar with a good bit of elementary Euclidean geometry since all of high school geometry falls into that category. *Advanced* Euclidean geometry is the geometry that was discovered later—it is geometry that was done after Euclid's death but is still built on Euclid's work. It is to be distinguished from *non-Euclidean geometry*, which is geometry based on axioms that are different from those used by Euclid. Throughout the centuries since Euclid lived, geometers have continued to develop Euclidean geometry and have discovered large numbers of interesting relationships. Their discoveries constitute advanced Euclidean geometry and are the subject matter of this text.

Many of the results of advanced Euclidean geometry are quite surprising. Most people who study them for the first time find the theorems to be amazing, almost miraculous, and value them for their aesthetic appeal as much as for their utility. I hope that users of this book will come to appreciate the elegance and beauty of Euclidean geometry and better understand why the subject has captivated the interest of so many people over the past two thousand years.

The book includes a study of the Poincaré disk model for hyperbolic geometry. Since this model is built within Euclidean geometry, it is an appropriate topic for study in a course on Euclidean geometry. Euclidean constructions, mostly utilizing inversions in circles, are used to illustrate many of the standard results of hyperbolic geometry.

Computer software

This is not the kind of textbook that neatly lays out all the facts you should know about advanced Euclidean geometry. Instead, it is meant to be a guide to the subject that leads you to discover both the theorems and their proofs for yourself. To fully appreciate the geometry presented here, it is essential that you be actively involved in the exploration and discovery process. Do not read the book passively, but diligently work through the explorations yourself as you read them.

The main tool used to facilitate active involvement and discovery is the software package GeoGebra. It enables users to explore the theorems of advanced Euclidean geometry, to discover many of the results for themselves, and to see the remarkable relationships with their own eyes.

The book consists mostly of exercises, tied together by short explanations. The user of the book should work through all the exercises while reading the book. That way he or she will be guided through the discovery process. Any exercise that is marked with a star (*) is meant to be worked on a computer, using GeoGebra, while the remaining exercises should be worked using pencil and paper. No prior knowledge of GeoGebra is assumed; complete instructions on how to use GeoGebra are included in Chapters 1 and 3.

GeoGebra is open source software that can be obtained free of charge from the website www.geogebra.org. That the software is free is important because it means that every student can have a copy. I believe it is essential that all students experience the discovery of geometric relationships for themselves. When expensive software packages are used, there is often only a limited number of copies available and not every student has access to one. Every student can have GeoGebra available all the time.

One of the best features of GeoGebra is how easy it is to use. Even a beginner can quickly produce intricate diagrams that illustrate complicated geometric relationships. Users soon learn to make useful tools that automate parts of the constructions. To ensure that every user of this book has the opportunity to experience that first hand, the reader is expected to produce essentially all the diagrams and illustrations. For that reason the number of figures in the text is kept to a minimum and no disk containing professionally-produced GeoGebra documents is supplied with the book.

GeoGebra is rapidly becoming the most popular and most widely used dynamic software package for geometry, but it is not the only one that can be used in conjunction with this text. Such programs as Geometer's Sketchpad, Cabri Geometry, Cinderella, and Geometry Expressions can also be utilized. The instructions that are included in Chapters 1 and 3 are specific to GeoGebra, but the rest of the book can be studied using any one of the programs mentioned.

Proof

A major accomplishment of the ancient Greeks was the introduction of logic and rigor into geometry. They *proved* their theorems from first principles and thus their results are more certain and lasting than are mere observations from empirical data. The logical, deductive aspect of geometry is epitomized in Euclid's *Elements* and proof continues to be one of the hallmarks of geometry to this day.

Until recently, all those who worked on advanced Euclidean geometry followed in Euclid's footsteps and did geometry by proving theorems, using only pencil and paper. Now that computer programs such as GeoGebra are available as tools, we must reexamine the place of proof in geometry. Some might expect the use of dynamic software to displace the deductive approach to geometry, but there is no reason the two approaches cannot enhance each other. I hope this book will demonstrate that proof and computer exploration can coexist comfortably in geometry and that each can support the other.

The exercises in this book will guide the student to use GeoGebra to explore and discover the statements of the theorems and then will go on to use GeoGebra to better understand the proofs of the theorems as well. At the end of this process of discovery the student should be able to write a proof of the result that has been discovered. In this way the student will come to understand the material to a depth that would not be possible if just computer exploration or just pencil and paper proof were used and should come to appreciate the fact that proof is an integral part of exploration, discovery, and understanding in mathematics.

Not only is proof an important part of the process by which we come to discover and understand geometric results, but the proofs also have a subtle beauty of their own. I hope that the experience of writing the proofs will help students to appreciate this aesthetic aspect of the subject as well.

In this text the word "verify" will be used to describe the kind of confirmation that is possible with GeoGebra. Thus to *verify* that the angle sum of a triangle is 180° will mean to use GeoGebra to construct a triangle, measure its three angles, calculate the sum of the measures, and then to observe that GeoGebra reports that the sum is always equal to 180° regardless of how the size and shape of the triangle are changed. On the other hand, to *prove* that the angle sum is 180° will mean to supply a written logical argument based on the axioms and previously proved theorems of Euclidean geometry.

Two ways to use this book

This book can be used as a manual for a computer laboratory that supplements a course in the foundations of geometry. The notation and terminology used here are consistent with *The Foundations of Geometry* [11], but this manual is designed to be used alongside any textbook on axiomatic geometry. The review chapter that is included at the beginning of the book establishes all the necessary terminology and notation.

A class that meets for one three-hour computer lab session per week should be able to lightly cover most of the text in one semester. When the book is used as a lab manual, Chapter 0 is not covered separately, but serves as a reference for notation, terminology, and statements of theorems from elementary Euclidean geometry. Most of the other chapters can be covered in one laboratory session each. The exceptions are Chapters 6 and 10, which are quite short and could be combined, and Chapter 11, which will require two or three sessions to cover completely.

A course that emphasizes Euclidean geometry exclusively will omit Chapter 14 and probably Chapter 13 as well, since the main purpose of Chapter 13 is to develop the tools that are needed for Chapter 14. On the other hand, most instructors who are teaching a course that covers non-Euclidean geometry will want to cover the last chapter; to do so it will probably be necessary to omit many of the applications of the Theorem of Menelaus. A thorough coverage of Chapter 14 will require more than one session.

At each lab session the instructor should assign an appropriate number of GeoGebra exercises, determined by the background of the students and the length of the laboratory session. It should be possible for students to read the short explanations during the session and work through the exercises on their own. A limited number of the written proofs can be assigned as homework following the lab session.

A second way in which to use the book is as a text for an inquiry-based course in advanced Euclidean geometry. Such a course would be taught in a modified Moore style in which the instructor does almost no lecturing, but students work out the proofs for themselves. A course based on these notes would differ from other Moore-style courses in the use of computer software to facilitate the discovery and proof phases of the process. Another difference between this course and the traditional Moore-style course is that students should be encouraged to discuss the results of their GeoGebra explorations with each other. Class time is used for student computer exploration and student presentations of solutions to exercises. The notes break down the proofs into steps of manageable size and offer students numerous hints. It is my experience that the hints and suggestions offered are sufficient to allow students to construct their own proofs of the theorems. The GeoGebra explorations form an integral part of the process of discovering the proof as well as the statement of the theorem. This second type of course would cover the entire book, including Chapter 0 and all the exercises in all chapters.

The preparation of teachers

The basic recommendation in *The Mathematical Education of Teachers* [2] is that future teachers take courses that develop a deep understanding of the mathematics they will teach. There are many ways in which to achieve depth in geometry. One way, for example, is to understand what lies beneath high school geometry. This is accomplished by studying the foundations of geometry, by examining the assumptions that lie at the heart of the subject, and by understanding how the results of the subject are built on those assumptions. Another way in which to achieve depth is to investigate what is built on top of the geometry that is included in the high school curriculum. That is what this course is designed to do.

One direct benefit of this course to future high school mathematics teachers is that those who take the course will develop facility in the use of GeoGebra. Dynamic geometry software such as GeoGebra will undoubtedly become much more common in the high school classroom in the future, so future teachers need to know how to use it and what it can do. In addition, software such as GeoGebra will likely lead to a revival of interest in advanced Euclidean geometry. When students learn to use GeoGebra they will have the capability to investigate geometric relationships that are more intricate than those studied in the traditional high school geometry course. A teacher who knows some advanced Euclidean geometry will have a store of interesting geometric results that can be used to motivate and excite students.

Do it yourself!

The philosophy of these notes is that students can and should work out the geometry for themselves. But students will soon discover that many of the GeoGebra tools they are asked to make in the exercises can be found on the world wide web. I believe students should be encouraged to make use of the mathematical resources available on the web, but that they also benefit from the experience of making the tools for themselves. Downloading a tool that someone else has made and using it is too passive an activity. Working through the constructions for themselves and seeing how the intricate constructions of advanced Euclidean geometry are based on the simple constructions from high school geometry will enable them to achieve a much deeper understanding than they would if they simply used ready-made tools.

I believe it is especially important that future high school mathematics teachers have the experience of doing the constructions for themselves. Only in this way do they come to know that they can truly understand mathematics for themselves and that they do not have to rely on others to work it out for them.

The question of whether or not to rely on tools made by others comes up most especially in the last chapter. There are numerous high-quality tools available on the web that can be used to perform constructions in the Poincaré disk. Nonetheless I think students should work through the constructions for themselves so that they clearly understand how the hyperbolic constructions are built on Euclidean ones. After they have built rudimentary tools of their own, they might want to find more polished tools on the web and add those to their toolboxes.

Acknowledgments

I want to thank all those who helped me develop this manuscript. Numerous Calvin College students and my colleague Chris Moseley gave useful feedback. Gerald Bryce and the following members of the MAA's Classroom Resource Materials Editorial Board read the manuscript carefully and offered many valuable suggestions: Michael Bardzell, Salisbury University; Diane Hermann, University of Chicago; Phil Mummert, Taylor University; Phil Straffin, Beloit College; Susan Staples, Texas Christian University; Cynthia Woodburn, Pittsburgh State University; and Holly Zullo, Carroll College. I also thank the members of the MAA publications department, especially Carol Baxter and Beverly Ruedi, for their help and for making the production process go smoothly. Finally, I thank my wife Patricia whose patient support is essential to everything I do.

Gerard A. Venema
venema@calvin.edu
April, 2013

Contents

Preface vii

0 A Quick Review of Elementary Euclidean Geometry 1
- 0.1 Measurement and congruence . . . 1
- 0.2 Angle addition . . . 2
- 0.3 Triangles and triangle congruence conditions . . . 3
- 0.4 Separation and continuity . . . 4
- 0.5 The exterior angle theorem . . . 5
- 0.6 Perpendicular lines and parallel lines . . . 5
- 0.7 The Pythagorean theorem . . . 7
- 0.8 Similar triangles . . . 8
- 0.9 Quadrilaterals . . . 9
- 0.10 Circles and inscribed angles . . . 10
- 0.11 Area . . . 11

1 The Elements of GeoGebra 13
- 1.1 Getting started: the GeoGebra toolbar . . . 13
- 1.2 Simple constructions and the drag test . . . 16
- 1.3 Measurement and calculation . . . 18
- 1.4 Enhancing the sketch . . . 20

2 The Classical Triangle Centers 23
- 2.1 Concurrent lines . . . 23
- 2.2 Medians and the centroid . . . 24
- 2.3 Altitudes and the orthocenter . . . 25
- 2.4 Perpendicular bisectors and the circumcenter . . . 26
- 2.5 The Euler line . . . 27

3 Advanced Techniques in GeoGebra 31
- 3.1 User-defined tools . . . 31
- 3.2 Check boxes . . . 33
- 3.3 The Pythagorean theorem revisited . . . 34

4 Circumscribed, Inscribed, and Escribed Circles 39
4.1 The circumscribed circle and the circumcenter 39
4.2 The inscribed circle and the incenter 41
4.3 The escribed circles and the excenters 42
4.4 The Gergonne point and the Nagel point 43
4.5 Heron's formula 44

5 The Medial and Orthic Triangles 47
5.1 The medial triangle 47
5.2 The orthic triangle 48
5.3 Cevian triangles 50
5.4 Pedal triangles 51

6 Quadrilaterals 53
6.1 Basic definitions 53
6.2 Convex and crossed quadrilaterals 54
6.3 Cyclic quadrilaterals 55
6.4 Diagonals 56

7 The Nine-Point Circle 57
7.1 The nine-point circle 57
7.2 The nine-point center 59
7.3 Feuerbach's theorem 60

8 Ceva's Theorem 63
8.1 Exploring Ceva's theorem 63
8.2 Sensed ratios and ideal points 65
8.3 The standard form of Ceva's theorem 68
8.4 The trigonometric form of Ceva's theorem 71
8.5 The concurrence theorems 72
8.6 Isotomic and isogonal conjugates and the symmedian point 73

9 The Theorem of Menelaus 77
9.1 Duality 77
9.2 The theorem of Menelaus 78

10 Circles and Lines 81
10.1 The power of a point 81
10.2 The radical axis 83
10.3 The radical center 84

11 Applications of the Theorem of Menelaus 85
11.1 Tangent lines and angle bisectors 85
11.2 Desargues' theorem 86
11.3 Pascal's mystic hexagram 88
11.4 Brianchon's theorem 90
11.5 Pappus's theorem 91
11.6 Simson's theorem 93
11.7 Ptolemy's theorem 96

11.8 The butterfly theorem 97

12 Additional Topics in Triangle Geometry **99**

12.1 Napoleon's theorem and the Napoleon point 99
12.2 The Torricelli point 100
12.3 van Aubel's theorem 100
12.4 Miquel's theorem and Miquel points 101
12.5 The Fermat point 101
12.6 Morley's theorem 102

13 Inversions in Circles **105**

13.1 Inverting points 105
13.2 Inverting circles and lines 107
13.3 Othogonality 108
13.4 Angles and distances 110

14 The Poincaré Disk **111**

14.1 The Poincaré disk model for hyperbolic geometry 111
14.2 The hyperbolic straightedge 113
14.3 Common perpendiculars 114
14.4 The hyperbolic compass 116
14.5 Other hyperbolic tools 117
14.6 Triangle centers in hyperbolic geometry 118

References **121**

Index **123**

About the Author **129**

0

A Quick Review of Elementary Euclidean Geometry

This preliminary chapter reviews basic results from elementary Euclidean geometry that will be needed in the remainder of the book. It assumes that readers have already studied elementary Euclidean geometry; the purpose of the chapter is to clarify which results will be used later and to introduce consistent notation. Readers who are using this book as a supplement to a course in the foundations of geometry can probably omit most of the chapter and simply refer to it as needed for a summary of the notation and terminology that are used in the remainder of the book.

The theorems in this chapter are to be assumed without proof; the entire chapter may be viewed as an extended set of axioms for the subject of advanced Euclidean geometry. The results in the exercises should be proved using the theorems stated in the chapter. All the exercises in the chapter are theorems that will be needed later.

We will usually refer directly to Euclid's *Elements* when we cite a result from elementary Euclidean geometry. Several current editions of the *Elements* are listed in the bibliography (see [4], [5], or [8]). The *Elements* are in the public domain and are freely available on the world wide web as well. Euclid's propositions are referenced by book number followed by the proposition number within that book. Thus, for example, Proposition III.36 refers to the 36th proposition in Book III of the *Elements*.

0.1 Measurement and congruence

For each pair of points A and B in the plane there is a nonnegative number AB, called the *distance* from A to B. The point B is *between* A and C if B is different from A and C and the distances are additive in the sense that $AB + BC = AC$. If B is between A and C, then A, B, and C are collinear. Conversely, if three distinct points are collinear, then one of them is between the other two.

The *segment* from A to B, denoted $\overline{AB}$, consists of A and B together with the points between A and B. The *length* of $\overline{AB}$ is the distance from A to B. Two segments $\overline{AB}$ and $\overline{CD}$ are *congruent*, written $\overline{AB} \cong \overline{CD}$, if they have the same length. There is also a *ray* $\overrightarrow{AB}$ and a *line* $\overleftrightarrow{AB}$ defined in the expected way—see Figure 0.1.

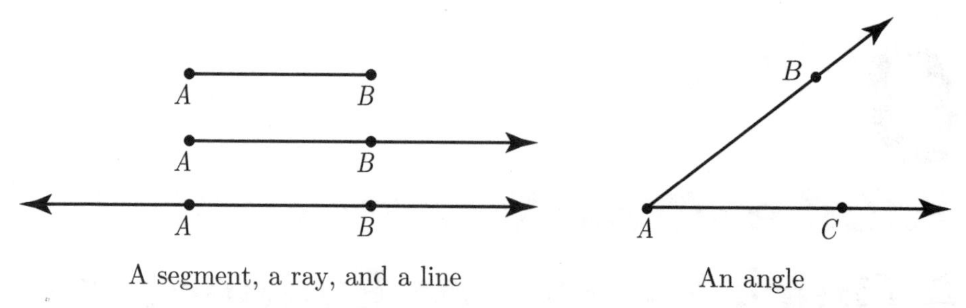

Figure 0.1. Basic geometric objects

For each triple of points A, B, and C with $A \neq B$ and $A \neq C$ there is an *angle*, denoted $\angle BAC$, that is defined by $\angle BAC = \overrightarrow{AB} \cup \overrightarrow{AC}$. The *measure* of the angle is a number $\mu(\angle BAC)$. We will always measure angles in degrees and assume that $0 \leq \mu(\angle BAC) \leq 180°$. The measure is $0°$ if the two rays $\overrightarrow{AB}$ and $\overrightarrow{AC}$ are equal; the measure is $180°$ if the rays are opposite; otherwise it is between $0°$ and $180°$. An angle is *acute* if its measure is less than $90°$, it is *right* if its measure equals $90°$, and it is *obtuse* if its measure is greater than $90°$. Two angles are *congruent* if they have the same measure.

0.2 Angle addition

Let A, B, and C be three noncollinear points. A point P is in the *interior* of $\angle BAC$ if P is on the same side of $\overleftrightarrow{AB}$ as C and on the same side of $\overleftrightarrow{AC}$ as B. The interior of $\angle BAC$ is defined provided $0° < \mu(\angle BAC) < 180°$. It is reasonable to define the interior of $\angle BAC$ to be the empty set in case $\mu(\angle BAC) = 0°$, but there is no good way to define the interior of an angle of measure $180°$.

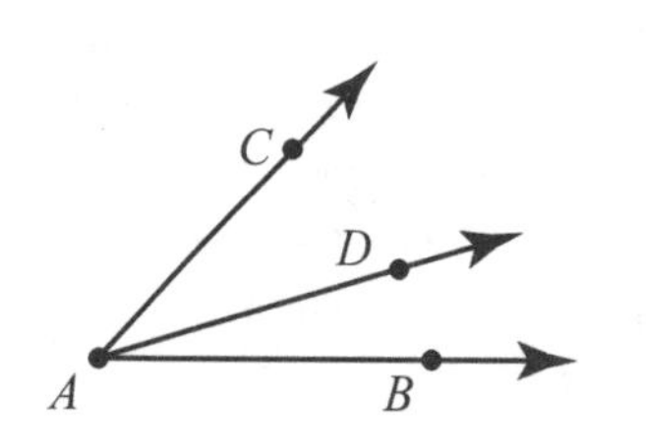

A ray $\overrightarrow{AD}$ is *between* rays $\overrightarrow{AB}$ and $\overrightarrow{AC}$ if D is in the interior of $\angle BAC$.

Angle Addition Postulate. *Let A, B, and C be three noncollinear points. The ray $\overrightarrow{AD}$ is* between *rays $\overrightarrow{AB}$ and $\overrightarrow{AC}$ if and only if $\mu(\angle BAC) = \mu(\angle BAD) + \mu(\angle DAC)$.*

There is an analogous theorem for angles of measure $180°$. Angles $\angle BAD$ and $\angle DAC$ form a *linear pair* if A, B, and C are collinear and A is between B and C.

Linear Pair Theorem. *If angles $\angle BAD$ and $\angle DAC$ form a linear pair, then $\mu(\angle BAD) + \mu(\angle DAC) = 180°$.*

Two angles whose measures add to $180°$ are called *supplementary angles* or *supplements*. The linear pair theorem asserts that if two angles form a linear pair, then they are supplements.

Angles $\angle BAC$ and $\angle DAE$ form a *vertical pair* (or are *vertical angles*) if rays $\overrightarrow{AB}$ and $\overrightarrow{AE}$ are opposite and rays $\overrightarrow{AC}$ and $\overrightarrow{AD}$ are opposite or if rays $\overrightarrow{AB}$ and $\overrightarrow{AD}$ are opposite and rays $\overrightarrow{AC}$ and $\overrightarrow{AE}$ are opposite.

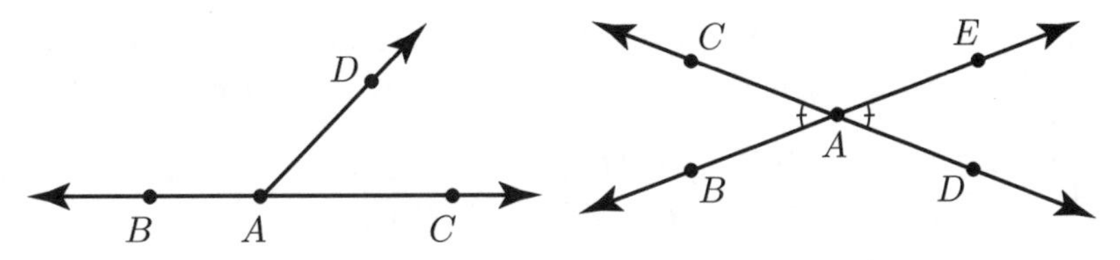

Figure 0.2. A linear pair and a vertical pair

Vertical Angles Theorem. *Vertical angles are congruent.*

The angle addition postulate and the linear pair theorem are not found in the *Elements* because Euclid did not use angle measure; instead he simply called two angles "equal" if, in our terminology, they are congruent. The vertical angles theorem is Euclid's Proposition I.15. The vertical angles theorem is a corollary of the linear pair theorem; for example, $\angle BAC \cong \angle DAE$ in the right half of Figure 0.2 because both are supplements of $\angle CAE$.

0.3 Triangles and triangle congruence conditions

The *triangle* with *vertices* A, B, and C consists of the points on the three segments determined by the three vertices; i.e., $\triangle ABC = \overline{AB} \cup \overline{BC} \cup \overline{AC}$.

The segments $\overline{AB}$, $\overline{BC}$, and $\overline{AC}$ are called the *sides* of the triangle $\triangle ABC$. Two triangles are *congruent* if there is a correspondence between the vertices of the first triangle and the vertices of the second triangle such that corresponding angles are congruent and corresponding sides are congruent.

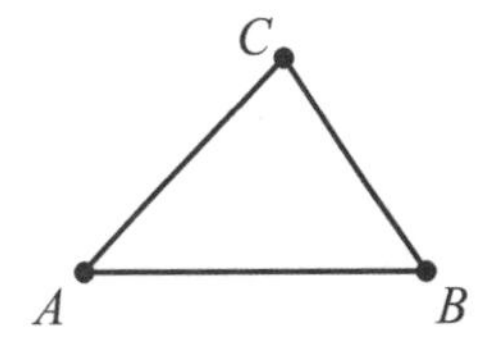

Notation. It is understood that the notation $\triangle ABC \cong \triangle DEF$ means that the two triangles are congruent under the correspondence $A \leftrightarrow D, B \leftrightarrow E$, and $C \leftrightarrow F$. The assertion that two triangles are congruent is really the assertion that there are six congruences, three angle congruences and three segment congruences. Specifically, $\triangle ABC \cong \triangle DEF$ means $\overline{AB} \cong \overline{DE}$, $\overline{BC} \cong \overline{EF}$, $\overline{AC} \cong \overline{DF}$, $\angle ABC \cong \angle DEF$, $\angle BCA \cong \angle EFD$, and $\angle CAB \cong \angle FDE$. In high school this is often abbreviated CPCTC (corresponding parts of congruent triangles are congruent).

If three parts of one triangle are congruent to the corresponding parts of another triangle, it is usually possible to conclude that the other three parts of the triangles are congruent as well. That is the content of the *triangle congruence conditions*.

Side-Angle-Side Theorem (SAS). *If $\triangle ABC$ and $\triangle DEF$ are triangles such that $\overline{AB} \cong \overline{DE}$, $\angle ABC \cong \angle DEF$, and $\overline{BC} \cong \overline{EF}$, then $\triangle ABC \cong \triangle DEF$.*

Euclid used his "method of superposition" to prove SAS (Proposition I.4), but it is usually taken to be a postulate in modern treatments of geometry. The next two results (ASA and AAS) are both contained in Euclid's Proposition I.26 and the third (SSS) is Euclid's Proposition I.8.

Angle-Side-Angle Theorem (ASA). *If $\triangle ABC$ and $\triangle DEF$ are triangles such that $\angle CAB \cong \angle FDE$, $\overline{AB} \cong \overline{DE}$, and $\angle ABC \cong \angle DEF$, then $\triangle ABC \cong \triangle DEF$.*

Angle-Angle-Side Theorem (AAS). *If $\triangle ABC$ and $\triangle DEF$ are triangles such that $\angle ABC \cong \angle DEF$, $\angle BCA \cong \angle EFD$, and $\overline{AC} \cong \overline{DF}$, then $\triangle ABC \cong \triangle DEF$.*

Side-Side-Side Theorem (SSS). *If $\triangle ABC$ and $\triangle DEF$ are triangles such that $\overline{AB} \cong \overline{DE}$, $\overline{BC} \cong \overline{EF}$, and $\overline{CA} \cong \overline{FD}$, then $\triangle ABC \cong \triangle DEF$.*

There is no side-side-angle condition, except in the special case in which the angle is a right angle.

Hypotenuse-Leg Theorem (HL). *If $\triangle ABC$ and $\triangle DEF$ are right triangles with right angles at the vertices C and F, respectively, $\overline{AB} \cong \overline{DE}$, and $\overline{BC} \cong \overline{EF}$, then $\triangle ABC \cong \triangle DEF$.*

Exercises

0.3.1. Use SAS to prove the following theorem (Euclid's Proposition I.5).

Isosceles Triangle Theorem. *If $\triangle ABC$ is a triangle and $\overline{AB} \cong \overline{AC}$, then $\angle ABC \cong \angle ACB$.*

0.3.2. Draw an example of two triangles that satisfy the SSA condition but are not congruent.

0.3.3. The *angle bisector* of $\angle BAC$ is a ray $\overrightarrow{AD}$ such that $\overrightarrow{AD}$ is between $\overrightarrow{AB}$ and $\overrightarrow{AC}$ and $\mu(\angle BAD) = \mu(\angle DAC)$. Recall that the distance from a point to a line is measured along a perpendicular from the point to the line. Prove the following theorem.

Pointwise Characterization of Angle Bisector. *A point P lies on the bisector of $\angle BAC$ if and only if P is in the interior of $\angle BAC$ and the distance from P to $\overleftrightarrow{AB}$ equals the distance from P to $\overleftrightarrow{AC}$.*

0.4 Separation and continuity

At the foundations of Euclidean geometry lie some profoundly deep assumptions regarding the way in which lines separate the plane and about the continuity of both circles and lines. These are the kinds of hypotheses that Euclid himself did not state explicitly, but used without comment. In modern treatments of geometry we strive to state all of our premises, so we think it is important to clarify what our assumptions are. In this section we state several such principles that are used repeatedly in Euclidean constructions. Following Euclid, we will not specifically refer to these results every time we use them.

The first statement is named for Moritz Pasch (1843–1930); it spells out how a line separates the plane into two (convex) subsets.

Pasch's Axiom. *Let $\triangle ABC$ be a triangle and let ℓ be a line such that none of the vertices A, B, and C lie on ℓ. If ℓ intersects $\overline{AB}$, then ℓ also intersects either $\overline{BC}$ or $\overline{AC}$ (but not both).*

The segment $\overline{BC}$ is called a *crossbar* for $\angle BAC$.

The second foundational principle codifies the intuitively "obvious" fact that it is impossible to leave the interior of a triangle without crossing one of the sides.

Crossbar Theorem. *If D is in the interior of $\angle BAC$, then there is a point G such that G lies on both $\overrightarrow{AD}$ and $\overline{BC}$.*

The next two statements are needed in our constructions to guarantee the existence of points of intersection. Since this is usually clear from the diagrams, we will assume them without comment. Euclid needed circle-circle continuity in the proof of his very first proposition, but did not think it necessary to state it as an explicit assumption.

Circle-Line Continuity. *If γ is a circle and ℓ is a line such that ℓ contains a point that is inside γ, then $\ell \cap \gamma$ consists of exactly two points.*

Circle-Circle Continuity. *Let α and β be two circles. If there exists a point of α that is inside β and there exists another point of α that is outside β, then $\alpha \cap \beta$ consists of exactly two points.*

0.5 The exterior angle theorem

The exterior angle theorem is an inequality regarding the angles in a triangle that is of fundamental importance in many of the proofs of elementary geometry. It is Euclid's Proposition I.16.

Let $\triangle ABC$ be a triangle. At each vertex of the triangle there is an *interior angle* and two *exterior angles*. The interior angle at A is the angle $\angle BAC$. The two angles $\angle CAD$ and $\angle BAE$ shown in Figure 0.3 are the exterior angles at A. The exterior angles at a vertex form a vertical pair and are therefore congruent.

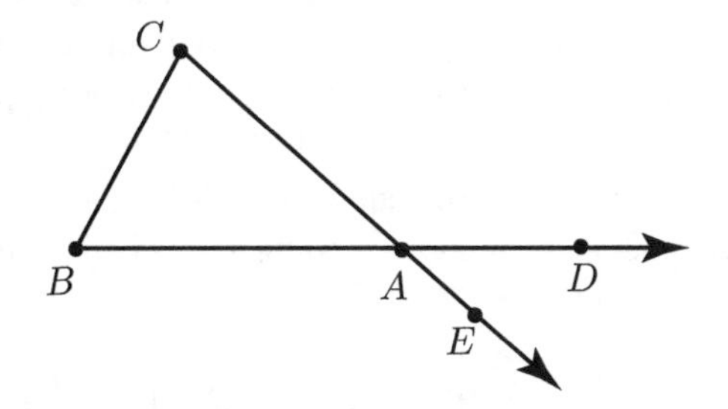

Figure 0.3. At each vertex there is one interior angle and there are two exterior angles

Exterior Angle Theorem. *The measure of an exterior angle for a triangle is strictly greater than the measure of either remote interior angle.*

0.6 Perpendicular lines and parallel lines

Two lines ℓ and m are *perpendicular*, written $\ell \perp m$, if they intersect at right angles. If ℓ is a line and P is a point, then there is exactly one line m such that P lies on m and $m \perp \ell$. The point at which m intersects ℓ is called the *foot* of the perpendicular from P

to ℓ. If P lies on ℓ, P itself is the foot of the perpendicular. The process of constructing the perpendicular m is called *dropping a perpendicular*—see Figure 0.4. Euclid proved that it is possible to construct the unique perpendicular with compass and straightedge (Proposition I.12).

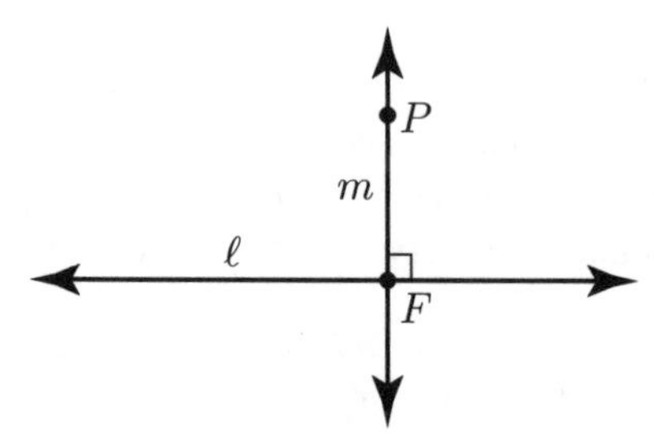

Figure 0.4. F is the foot of the perpendicular from P to ℓ

Two lines ℓ and m in the plane are *parallel*, written $\ell \parallel m$, if they do not intersect. It is the existence and uniqueness of parallels that distinguishes Euclidean geometry from non-Euclidean geometries. The Euclidean parallel property is stated most succinctly in the following postulate.

Playfair's Postulate. *For every line ℓ and for every point P that does not lie on ℓ there exists exactly one line m such that P lies on m and $m \parallel \ell$.*

In the presence of the other axioms of geometry, Playfair's postulate is equivalent to Euclid's fifth postulate. The next two theorems relate parallelism to angle congruence. They are a standard part of high school geometry and are also Propositions I.27 and I.29 in Euclid. It is in the proof of Proposition I.29 (the theorem we call the converse to the alternate interior angles theorem) that Euclid first uses his fifth postulate.

Let ℓ and ℓ' denote two lines in the plane. A *transveral* for the two lines is a line t such that t intersects ℓ and ℓ' in distinct points. The transversal makes a total of eight angles with the two lines—see Figure 0.5. The two pairs $\{\angle ABB', \angle BB'C'\}$ and $\{\angle A'B'B, \angle B'BC\}$ are called *alternate interior angles*. The angles $\{\angle ABB', \angle A'B'B''\}$ are *corresponding angles*. There are three other pairs of corresponding angles defined in the obvious way.

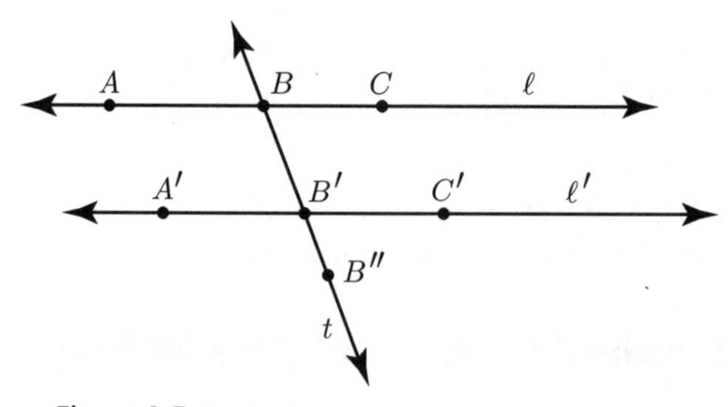

Figure 0.5. Angles formed by two lines and a transversal

Alternate Interior Angles Theorem. *If ℓ and ℓ' are two lines cut by a transversal t in such a way that a pair of alternate interior angles is congruent, then ℓ is parallel to ℓ'.*

Converse to the Alternate Interior Angles Theorem. *If two parallel lines are cut by a transversal, then both pairs of alternate interior angles are congruent.*

Exercises

0.6.1. Prove the following theorem (Euclid's Proposition I.28) and its converse.

> **Corresponding Angles Theorem.** *If ℓ and ℓ' are lines cut by a transversal t in such a way that two corresponding angles are congruent, then ℓ is parallel to ℓ'.*

0.6.2. Prove the following theorem (Euclid's Proposition I.32).

> **Angle Sum Theorem.** *For every triangle, the sum of the measures of the interior angles of the triangle is* $180°$.
>
> Hint: Let $\triangle ABC$ be a triangle. Draw a line through C that is parallel to the line through A and B. Then apply the converse to the alternate interior angles theorem.

0.6.3. The *perpendicular bisector* of a segment $\overline{AB}$ is a line ℓ such that ℓ intersects $\overline{AB}$ at its midpoint and $\ell \perp \overline{AB}$. Prove the following theorem.

> **Pointwise Characterization of Perpendicular Bisector.** *A point P lies on the perpendicular bisector of $\overline{AB}$ if and only if $PA = PB$.*

0.7 The Pythagorean theorem

The Pythagorean theorem is probably the most famous theorem in all of geometry; it is the one theorem that every high school student remembers. For Euclid it was the culmination of Book I of the *Elements*. The theorem is named for Pythagoras of Samos who lived from about 569 to 475 BC. Few details about the life of Pythagoras are known, so it is difficult to determine whether Pythagoras really did prove the theorem that bears his name or what kind of proof he might have used.

Notation. Let $\triangle ABC$ be a triangle. It is standard to use lower case letters to denote the lengths of the sides of the triangle: $a = BC, b = AC$, and $c = AB$.

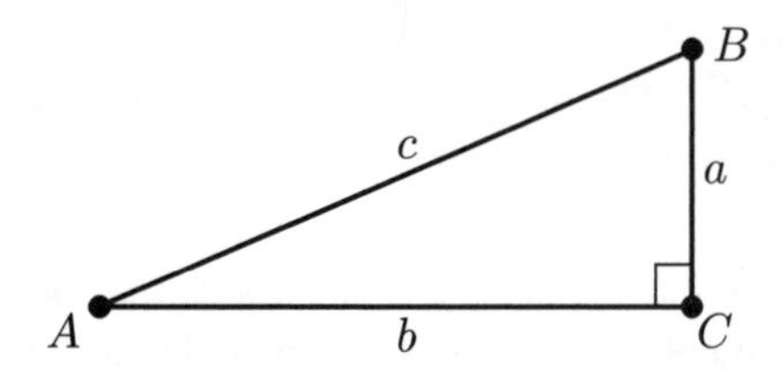

Figure 0.6. Standard right triangle notation

Pythagorean Theorem. *If $\triangle ABC$ is a right triangle with right angle at vertex C, then $a^2 + b^2 = c^2$.*

Euclid gave two kinds of proofs of the Pythagorean theorem; the first one based on area and then later another based on similar triangles.

0.8 Similar triangles

The similar triangles theorem is one of the most useful in elementary Euclidean geometry. Euclid did not prove it, however, until Book VI of the *Elements*. (The similar triangles theorem is Euclid's Proposition VI.4.) The reason he waited so long is that the ancient Greeks had trouble dealing with the irrational ratios that can arise when similar triangles are compared. It is believed that Eudoxus of Cnidus (408–305 BC) was the first to give a complete proof of the theorem.

Triangles $\triangle ABC$ and $\triangle DEF$ are *similar* if $\angle ABC \cong \angle DEF$, $\angle BCA \cong EFD$, and $\angle CAB \cong \angle FDE$. Write $\triangle ABC \sim \triangle DEF$ if $\triangle ABC$ is similar to $\triangle DEF$. As with congruence of triangles, the order in which the vertices are listed is significant.

Similar Triangles Theorem. *If $\triangle ABC$ and $\triangle DEF$ are two triangles such that $\triangle ABC \sim \triangle DEF$, then*

$$\frac{AB}{AC} = \frac{DE}{DF}.$$

Exercises

0.8.1. Prove the following theorem. It is a special case of the parallel projection theorem [11, Theorem 7.3.1] and will prove to be useful later.

Euclid's Proposition VI.2. *Let $\triangle ABC$ be a triangle, and let D and E be points on the sides $\overline{AB}$ and $\overline{AC}$, respectively. Then $\overleftrightarrow{DE} \parallel \overleftrightarrow{BC}$ if and only if $AD/AB = AE/AC$.*

Hint for converse: Assume $AD/AB = AE/AC$. Let ℓ be the line through D such that $\ell \parallel \overleftrightarrow{BC}$. Use Pasch's axiom to prove that there is a point E' where ℓ intersects $\overline{AC}$. Prove that $E' = E$.

0.8.2. Prove the following theorem (Euclid's Proposition VI.6).

SAS Similarity Criterion. *If $\triangle ABC$ and $\triangle DEF$ are two triangles such that $\angle CAB \cong \angle FDE$ and $AB/AC = DE/DF$, then $\triangle ABC \sim \triangle DEF$.*

Hint: If $AB = DE$, the proof is easy. Otherwise it may be assumed that $AB > DE$ (explain). Choose a point B' between A and B such that $AB' = DE$ and let m be the line through B' that is parallel to $\overleftrightarrow{BC}$. Prove that m intersects $\overline{AC}$ in a point C' such that $\triangle AB'C' \cong \triangle DEF$.

0.9 Quadrilaterals

Four points A, B, C, and D such that no three are collinear determine a *quadrilateral,* which we will denote by $\square ABCD$. Specifically,

$$\square ABCD = \overline{AB} \cup \overline{BC} \cup \overline{CD} \cup \overline{DA}.$$

It is usually assumed that the segments $\overline{AB}$, $\overline{BC}$, $\overline{CD}$, and $\overline{DA}$ intersect only at their endpoints, but we will relax that requirement later.

The four segments are called the *sides* of the quadrilateral and the points A, B, C, and D are called the *vertices* of the quadrilateral. The sides $\overline{AB}$ and $\overline{CD}$ are called *opposite sides* of the quadrilateral as are the sides $\overline{BC}$ and $\overline{AD}$. The segments $\overline{AC}$ and $\overline{BD}$ are the *diagonals* of the quadrilateral. Two quadrilaterals are *congruent* if there is a correspondence between their vertices so that all four corresponding sides are congruent and all four corresponding angles are congruent.

There are several special kinds of quadrilaterals that have names. A *trapezoid* is a quadrilateral in which at least one pair of opposite sides is parallel. A *parallelogram* is a quadrilateral in which both pairs of opposite sides are parallel. It is obvious that every parallelogram is a trapezoid, but not vice versa. A *rhombus* is a quadrilateral in which all four sides are congruent. A *rectangle* is a quadrilateral in which all four angles are right angles. A *square* is a quadrilateral that is both a rhombus and a rectangle.

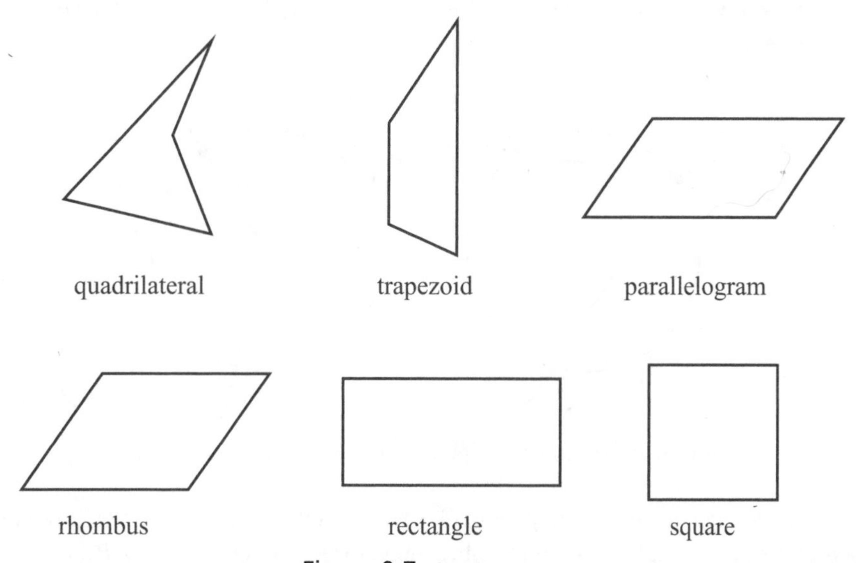

Figure 0.7. Quadrilaterals

Exercises

0.9.1. Prove the following theorem.

Euclid's Proposition I.34. *The opposite sides of a parallelogram are congruent.*

Hint: Draw a diagonal and use ASA.

0.9.2. Prove that the diagonals of a parallelogram bisect each other; i.e., if $\square ABCD$ is a parallelogram, then the diagonals $\overline{AC}$ and $\overline{BD}$ intersect in a point E and E is the midpoint of both diagonals.

0.10 Circles and inscribed angles

Let r be a positive number and let O be a point. The *circle* with *center* O and *radius* r is defined by $\mathcal{C}(O, r) = \{P \mid OP = r\}$. The *diameter* of the circle is $d = 2r$. While the radius of a circle is usually thought of as a number, it is often convenient to refer to one of the segments $\overline{OP}$, $P \in \mathcal{C}(O, r)$, as a radius of the circle $\mathcal{C}(O, r)$. In the same way, a segment $\overline{PQ}$ such that P and Q lie on the circle and $O \in \overline{PQ}$ is called a diameter of $\mathcal{C}(O, r)$.

Let γ be a circle and let P be a point on γ. A line t is *tangent to* γ *at* P if $t \cap \gamma = \{P\}$.

Tangent Line Theorem. *Let $\gamma = \mathcal{C}(O, r)$ be a circle and let ℓ be a line that intersects γ at P. Then ℓ is tangent to γ at P if and only if $\ell \perp \overleftrightarrow{OP}$.*

Let $\gamma = \mathcal{C}(O, r)$ be a circle. An *inscribed angle* for γ is an angle of the form $\angle PQR$, where P, Q, and R all lie on γ. The *arc intercepted* by the inscribed angle $\angle PQR$ is the set of points on γ that lie in the interior of $\angle PQR$.

Inscribed Angle Theorem. *If two inscribed angles intercept the same arc, then the angles are congruent.*

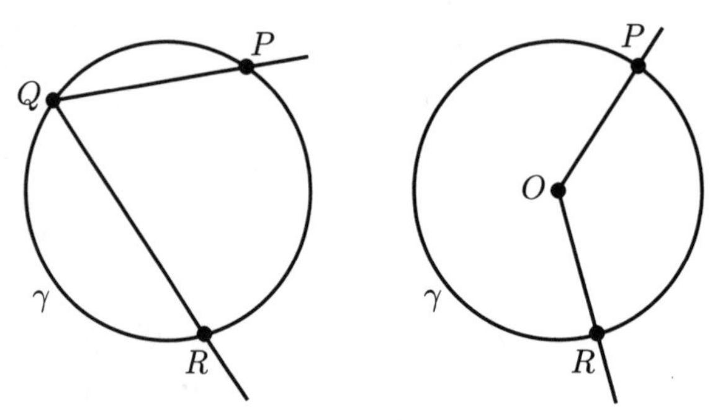

Figure 0.8. An inscribed angle and the corresponding central angle

A *central angle* for the circle $\gamma = \mathcal{C}(O, r)$ is an angle of the form $\angle POR$, where P, and R all lie on γ. If $\angle PQR$ is an inscribed angle for $\gamma = \mathcal{C}(O, r)$, then $\angle POR$ is called the *corresponding central angle*. (See Figure 0.8.)

Central Angle Theorem. *The measure of an inscribed angle is one half the measure of the corresponding central angle (provided the measure of the inscribed angle is less than or equal to* $90°$*).*

If we allowed angles of measure greater than $180°$ it would not be necessary to add the parenthetical provision at the end of the central angle theorem. The central angle theorem is Euclid's Proposition III.20 and the inscribed angle theorem is Euclid's Proposition III.21.

Exercises

0.10.1. Prove the following theorem. It is Euclid's Proposition III.3.

Secant Line Theorem. *If $\gamma = \mathcal{C}(O, r)$ is a circle and ℓ is a line that intersects γ at distinct points P and Q, then O lies on the perpendicular bisector of the chord $\overline{PQ}$.*

0.10.2. Prove the following theorem.

External Tangents Theorem. *If $\gamma = \mathcal{C}(O, r)$ is a circle and ℓ and m are two non-parallel lines that are tangent to γ at the points P and Q, and A is the point of intersection of ℓ and m, then $PA = QA$.*

0.10.3. The following theorem can be viewed as a special case of the inscribed angle theorem. Give a proof that does not use the inscribed angle theorem. The theorem is named for Thales of Miletus (624–547 BC); it is Euclid's Proposition III.31.

Thales' Theorem. *If the vertices of $\triangle ABC$ lie on a circle and $\overline{AB}$ is a diameter of that circle, then $\angle ACB$ is a right angle.*

Hint: Let O be the midpoint of $\overline{AB}$. Observe that $AO = BO = CO$ and apply the isosceles triangle theorem along with the angle sum theorem.

0.10.4. Prove the following theorem.

Converse to Thales' Theorem. *If $\angle ACB$ is a right angle, then the vertices of $\triangle ABC$ lie on a circle and $\overline{AB}$ is a diameter of that circle.*

Hint: Again let O be the midpoint of $\overline{AB}$. There is a point C' such that C' lies on $\overrightarrow{OC}$ and $OC' = OA$. Prove that $C = C'$.

0.10.5. Use the angle sum theorem and the linear pair theorem to prove the following special case of the central angle theorem: *If $\triangle ABC$ is a right triangle with right angle at C and O is the midpoint of $\overline{AB}$, then $\mu(\angle BOC) = 2\mu(\angle BAC)$.* This result will be used repeatedly in later chapters.

0.11 Area

A *polygon* is a generalization of triangle and quadrilateral. A polygon P has a finite set of *vertices* $A_1, A_2, \ldots, A_n$. The polygon is defined by

$$P = \overline{A_1A_2} \cup \overline{A_2A_3} \cup \cdots \cup \overline{A_{n-1}A_n} \cup \overline{A_nA_1}.$$

The segments $\overline{A_1A_2}$, etc., are called the *sides* of the polygon. The sides of a polygon are one-dimensional and have no area. Corresponding to each polygon in the plane there is a *region*, which consists of the points of the polygon itself together with the points inside it. It is the region that is two-dimensional and has area.

For each polygonal region in the plane there is a nonnegative number called the *area* of the region. The area of a region R is denoted by $\alpha(R)$. The area of a triangular region is

Figure 0.9. A polygon and the associated polygonal region

given by the familiar formula

$$\text{area} = (1/2)\ \text{base} \times \text{height}.$$

The other important property of area is that it is *additive*, which means that the area of a region that is the union of two nonoverlapping subregions is the sum of the areas of the subregions.

Exercise

0.11.1. Prove that the area of a triangle $\triangle ABC$ is given by

$$\alpha(\triangle ABC) = \frac{1}{2} AB \cdot AC \cdot \sin(\angle BAC).$$

1

The Elements of GeoGebra

The tool we will use to facilitate our exploration of advanced Euclidean geometry is the computer program GeoGebra. No prior knowledge of the software is assumed; this chapter takes a quick tour of GeoGebra and provides all the information you need to get started. The basic commands that are explained in the chapter are enough for you to begin creating sketches of your own. More advanced GeoGebra techniques, such as the creation of custom tools and the use of check boxes, will be discussed in Chapter 3.

GeoGebra is open source software that you can obtain free of charge. If you have not already done so, you should download a copy from the website www.geogebra.org and install it on your computer. Version 4.0 was used in the preparation of this book. There were some fairly significant changes from version 3 to 4 and there may be further modifications to various commands in later versions of the software.

Even though this chapter contains all the information you need to get started working with GeoGebra, you may still find it convenient to refer to other sources for particulars. Complete documentation of all features of GeoGebra is available on the GeoGebra website at wiki.geogebra.org/en/.

Throughout this chapter and the remainder of the book, exercises that are to be worked using GeoGebra are marked with *. All other exercises are traditional pencil-and-paper exercises.

1.1 Getting started: the GeoGebra toolbar

GeoGebra is dynamic software that is meant to be used in the teaching and learning of mathematics. GeoGebra combines geometry, algebra, and statistics in one package. Since we will be using it exclusively for geometry, you will usually want to hide the algebra and spreadsheet views and just work in the graphics view. The geometry we do will be mostly coordinate-free, so you should hide the coordinate axes as well. The View menu is used to customize the user interface. When you are satisfied with the appearance of your GeoGebra window, click Save Settings in the Settings box that is found under the Options menu. From then on your copy of GeoGebra will open with the appearance you have chosen.

All the tools you need to make sketches are found in the GeoGebra *toolbar* (Figure 1.1). When you click on the icon for one of the tools, the tool is activated and its name appears to the right of the toolbar. The tools are simple to use and the names are mostly self-explanatory. Each tool has several variants; they appear in a pull-down menu when you click on the ∇ in the lower right hand corner of the tool's icon. You can use the Customize Toolbar command under the Tools menu to change the appearance of the toolbar.

Figure 1.1. The GeoGebra toolbar

Here is a short description of what the first three tools do. You should try out each of the tools as you read the explanation of how it works. Other tools will be introduced later.

Move. The Move tool is used to select an object. Click on the icon to activate the tool and then click on an object in your sketch. You can now drag and drop the object.

New Point. The New Point tool creates a new point wherever you click in the graphics window.

Line tools. The line tools are used to create lines, rays, and segments (Figure 1.2). For example, to create a line you choose Line Through Two Points and then click on two points. For Ray Through Two Points, the first point chosen is the endpoint of the ray. To use Segment with Given Length from Point you choose one endpoint and then a dialog box pops up in which you enter the numerical value of the length. The segment will initially point to the right, but you can use the Move tool to relocate the second endpoint.

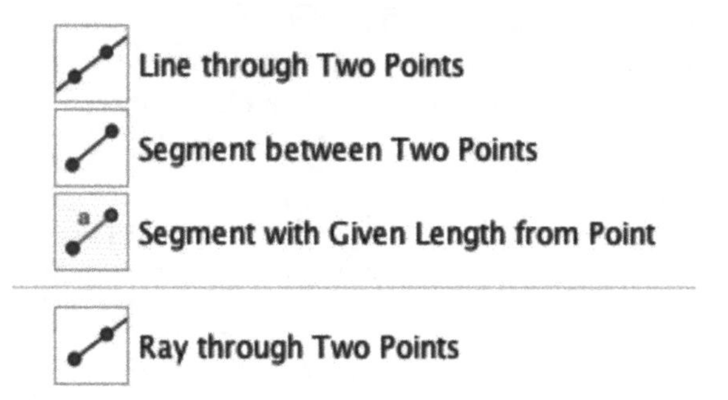

Figure 1.2. The GeoGebra line tools

Other point tools. There are several variations on the point tool (Figure 1.3). The Point on Object tool is used to create a point that is confined to a previously constructed object (such a segment or line). Activate the tool by clicking the icon and then click where you want to create a point on the object. The same result can be achieved with the basic Point tool; just watch for the object to become highlighted before you click to create the point. The Intersect Two Objects tool introduces the point at which two one-dimensional objects (such as lines and circles) intersect. To use it, you click on one of the objects and then the other. We will refer to this process as *marking* the point of intersection. The tool Midpoint or Center locates the midpoint or center of an object. For example, you can locate the

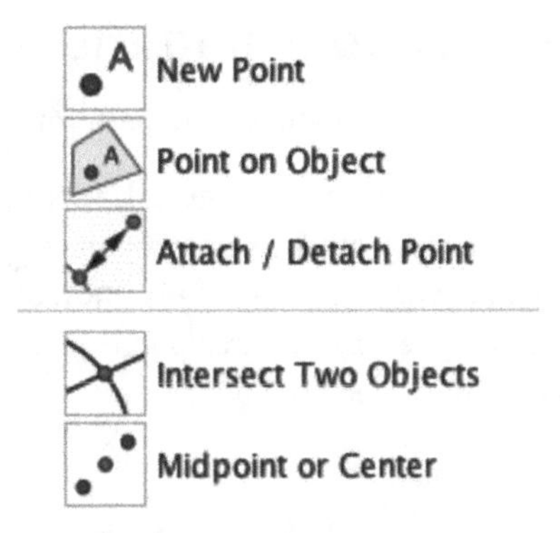

Figure 1.3. The GeoGebra point tools

midpoint of a segment by either clicking on the two endpoints in turn or by clicking on the segment itself.

Free and dependent objects. The objects you create with the tools described above are related to each other in various ways. For example, if you use the New Point tool to create a point somewhere in a blank window, that point is *free* in the sense that there is no restriction on where it can be moved with the Move tool. On the other hand, if you first construct a line segment and then use the Point on Object tool to construct a point on the segment, the result is not a free point but a *point on an object*. The point can be moved freely within the segment, but it cannot be moved off the segment. We will refer to such a point as a *movable point*. It has a limited range of motion available to it; it can move along the segment, but not off the segment. If you now use the Midpoint tool to construct the midpoint of the segment, the midpoint is completely determined by the (endpoints of the) segment and cannot be moved at all. It is a *dependent* object. It will move when the endpoints of the segment are moved, but it has no freedom whatsoever to move on its own. There is an Attach / Detach Point tool that can be used to establish or break dependency relationships between existing objects. The Object Properties box, which is found in the Edit menu, supplies a limited amount of information on the relationships between various objects in a sketch.

Exercises

*1.1.1. Construct two free points and then construct the line segment joining them. Construct a movable point on the segment and the midpoint of the segment. Observe what happens when you try to use the Move tool to move each of the four points you have constructed. Examine the Objects Properties box to familiarize yourself with how the relationships between these objects are described there.
Hint: Look for the Definition of the object.

*1.1.2. Use the Line Through Two Points tool to construct two intersecting lines. Mark the point of intersection. Observe what happens when you attempt to move each of the five points in your diagram.

1.2 Simple constructions and the drag test

The next three sets of tools in the toolbar allow us to perform simple constructions.

More line tools. The tools in this menu (Figure 1.4) are used to construct lines that are dependent objects, determined by other lines and points. For example, to drop a perpendicular from a point to a line, you activate the Perpendicular Line tool and then select the point and line (in either order). Click on three points that determine an angle in order to construct the angle bisector; the second point is the vertex of the angle.

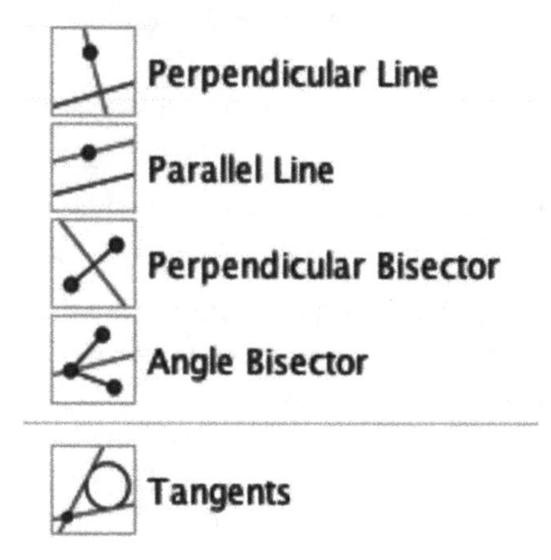

Figure 1.4. GeoGebra tools for constructing lines

Polygon tools. There are two ways to construct a polygon. One way is to use the Segment Between Two Points tool to construct the sides of the polygon one at a time. You can also construct a polygon with the Polygon tool. To do so, activate the tool and then click on the vertices of the polygon in order. The polygon will be complete when you close it by clicking on the first vertex a second time.

There is an important difference between the polygons constructed in these two ways. The first polygon is the union of a finite number of line segments and is, therefore, one-dimensional. The second method produces a two-dimensional object that includes the segments on the sides and the interior points as well. In other words, the first method produces what in elementary geometry we are accustomed to calling a polygon while the second method produces a polygonal region. This distinction will occasionally be important; for example, if we want to measure the area of a triangle we must construct it using the Polygon tool.

Circle tools. For now we need only the first of these tools, Circle with Center through Point. To use it, activate the tool and then click on two points. The result is a circle that contains the second point and has the first point as center.

Example. An *equilateral* triangle is a triangle in which all three sides have equal lengths. Euclid's very first proposition asserts that for every segment $\overline{AB}$, there is a point C such that $\triangle ABC$ is equilateral. His construction nicely illustrates how the tools we have described work.

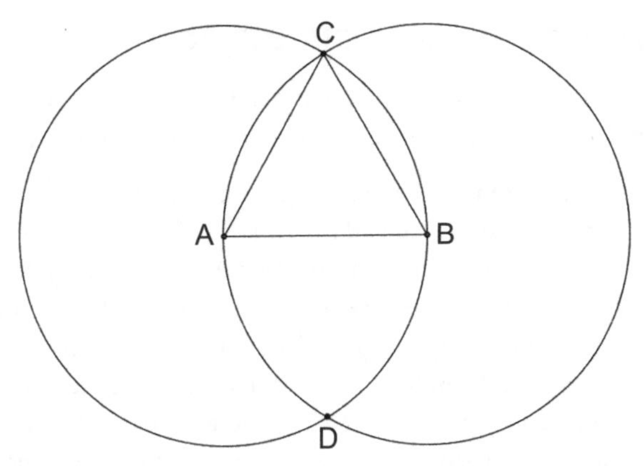

Figure 1.5. GeoGebra diagram of Euclid's construction

Begin by using the New Point tool to construct two points A and B and then use the Segment between Two Points tool to construct the segment $\overline{AB}$. Next use the Circle with Center through Point tool to construct the circle through B with center A and the circle through A with center B. Use the Intersect Two Objects to mark the points C and D of intersection of the circles. Finally, construct the segments $\overline{AC}$ and $\overline{BC}$. The resulting sketch is shown in Figure 1.5. □

The drag test. It appears from the diagram that $\triangle ABC$ is equilateral. In order to convince yourself that it is (and doesn't just happen to look equilateral because of the particular placement of A and B), use the Move tool to drag the two points A and B around with the mouse and watch what happens to $\triangle ABC$. The triangle in your diagram should continue to look equilateral. If it does, we say that the construction has passed the *drag test*. (It is only necessary to drag A and B because everything else in the diagram is dependent on them.) The drag test does not prove that the triangle is equilateral, but it is a good rough-and-ready way to check that the construction is probably succeeding.

Exercises

*1.2.1. Perform the following constructions to familiarize yourself with the tools described in the section. Use the drag test to check that your constructions are successful.

(a) Construct a line ℓ and a point P that does not lie on ℓ. Construct the line through P that is parallel to ℓ. Construct the line through P that is perpendicular to ℓ.

(b) Construct a segment and the perpendicular bisector of the segment.

(c) Construct an angle $\angle BAC$ and the bisector of $\angle BAC$.
Hint: An angle is the union of two rays with the same endpoint.

(d) Construct a circle γ, a movable point Q on γ, and the line that is tangent to γ at Q.

*1.2.2. Construct a segment $\overline{AB}$ and then construct a square that has $\overline{AB}$ as a side. Verify that your construction passes the drag test.
Hint: Construct a perpendicular line at A. Then construct a circle with center A and radius AB. Mark one of the points at which the perpendicular line intersects the circle. This is a third vertex of the square.

1.3 Measurement and calculation

The geometric quantities of distance (or length), angle measure, and area can all be measured in GeoGebra and assigned numerical values. We can then do calculations with them.

Measuring tools. All of GeoGebra's measuring tools may be found under the angle tool (Figure 1.6). Angle measures an angle in degrees. The angle is always measured in the counterclockwise direction, so the value of the measure will be a number in the range 0° to 360°. Distance or Length measures either the distance between two points or the length of a segment. (GeoGebra uses $\overline{AB}$ to denote the distance from A to B, while we use $\overline{AB}$ to denote the segment from A to B.) Area measures the area of a polygon. (Recall the comment about area and polygons in §1.2.) Angle with Given Size is the odd tool out in this group of tools; it does not measure anything, but instead constructs an angle of specified measure.

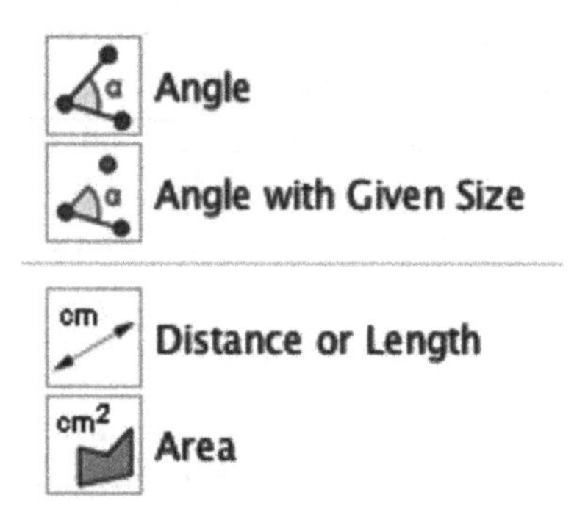

Figure 1.6. GeoGebra's measurement tools

Calculation. Calculations are done using the Input Bar in the Algebra View; to open it, choose Algebra View and Show Input Bar under the View menu. When you open the Algebra View you should see a list of all the objects in the current sketch, divided into Free Objects and Dependent Objects. As you measure objects in your sketch, the measurements will appear in the Algebra View and will automatically be assigned names. To calculate a quantity that depends on the measurements, you type the formula in the input bar.

Example. To illustrate how measurement and calculation work, we will verify the formula in the conclusion of the Pythagorean theorem. Begin by constructing two points A and B and the segment $\overline{AB}$. Then construct the line that is perpendicular to $\overleftrightarrow{AB}$ at B and create a new point C on the perpendicular. Finally, construct the segments $\overline{AC}$ and $\overline{BC}$. (The segment $\overline{BC}$ will be right on top of the line $\overleftrightarrow{BC}$, so it won't be separately visible. It should be constructed anyway so that it can be measured.) Now use the Distance or Length tool to measure the lengths of the three segments $\overline{AB}$, $\overline{BC}$ and $\overline{AC}$. The lengths should appear in the Geometry View as in the right half of Figure 1.7.

Before you can do any calculation you must open the Algebra View and activate the Input Bar. When you open the Algebra View you will see something like Figure 1.7. The only free objects in the sketch are the points A and B; the point C is a movable point on the perpendicular line. Notice that GeoGebra has named the sides of the triangle a, c, and d and that the perpendicular line is named b (assuming you constructed the objects in exactly

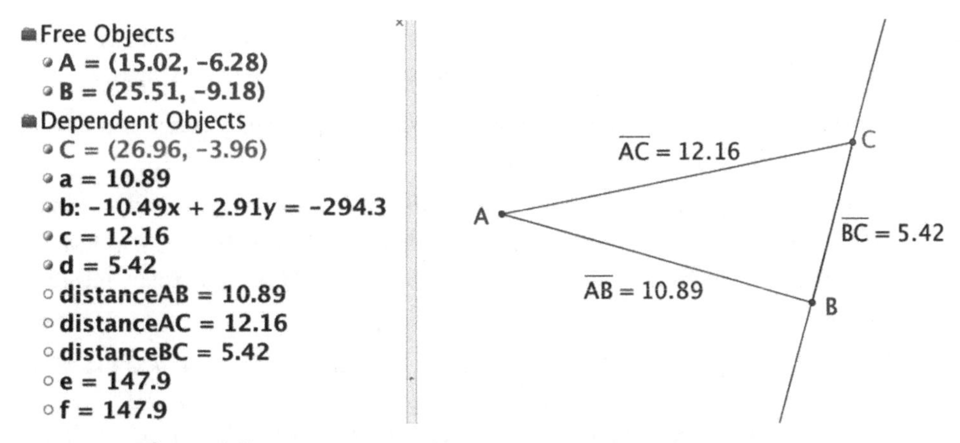

Figure 1.7. The Algebra and Geometry views of the example calculation

the order described in the example). We will verify that $a^2 + d^2 = b^2$. Type $a^2 + d^2$ in the Input Bar and press enter. The new quantity is assigned the name e. Then type b^2 in the Input Bar and press enter again. (To insert mathematical symbols, such as the superscript 2, you can either use the palette that appears when you click on the α at the right hand end of the Input Bar or you can use LaTeX commands, such as `a^2 + d^2`.) Notice that e and f are equal. Perform the drag test by moving B and by moving C on the perpendicular line. You should observe that e and f remain equal no matter where B and C are located. □

Exercises

*1.3.1. Go back to the equilateral triangle you constructed in §1.2. Measure the lengths of the sides of the triangle; observe that they are equal and remain equal when the drag test is applied. Measure the interior angles in the triangle and observe that the angles also have equal measure. (Be careful to choose the vertices in the correct order so that you measure the interior angles and not the outside angles of the triangle.)

*1.3.2. Construct a pair of parallel lines. Construct a movable point A on one line and construct points B and C on the other line. Construct the triangle $\triangle ABC$ and measure its area. Now drag the point A and observe that the area remains constant. (Make certain that the point moves but that the line does not.) Can you explain this in terms of the usual formula for the area of a triangle?

*1.3.3. Construct a triangle, measure each of the interior angles, and calculate the sum of their measures. Observe that the sum remains constant when the vertices of the triangle are dragged about in the plane.
Hint: GeoGebra will use Greek letters to denote the measures of the angles. You can find Greek characters in the palette that appears when you click on the small box marked α at the right-hand end of the Input Bar.

*1.3.4. Construct a triangle and an exterior angle for the triangle. Measure the exterior angle and compare it with the measures of the two remote interior angles. What is the exact relationship between the measure of the exterior angle and the measures of the two remote interior angles? Verify your answer by calculating the sum of the measures of the two remote interior angles. Move the vertices of the triangle around and make sure that the relationship you discovered is correct for triangles of all shapes.

1.3.5. State and prove the theorem you discovered in the preceding exercise. It should improve on the exterior angle theorem.

*1.3.6. Construct a circle γ and three points A, B, and C on γ. Measure $\angle BAC$. Then drag A, while keeping B and C fixed, and observe what happens to $\mu(\angle BAC)$. How many different values are assumed? How are they related? Can you explain this relationship?

*1.3.7. Construct a quadrilateral $\square ABCD$ and then construct the midpoints of the sides. Label the midpoints E, F, G, H and construct the quadrilateral $\square EFGH$. The new quadrilateral is called the *midpoint quadrilateral* for $\square ABCD$. Regardless of the shape of the original quarilateral, the midpoint quadrilateral always has a special property. What special kind of quadrilateral is $\square EFGH$? Drag the points A, B, C, and D around in the plane to change the shape of $\square ABCD$ to test your conjecture. Modify your conjecture as necessary.

*1.3.8. Construct a quadrilateral, its diagonals, and the midpoints of the diagonals. Experiment with the quadrilateral to determine answers to the following questions. (It might be helpful to review the names of the various kinds of quadrilaterals defined on page 9 before you begin.)

(a) For which quadrilaterals do the diagonals bisect each other? (Two segments *bisect each other* if they intersect at their midpoints.)

(b) For which quadrilaterals do the diagonals bisect each other and have the same length?

(c) For which quadrilaterals do the diagonals bisect each other and intersect at right angles?

(d) For which quadrilaterals do the diagonals bisect each other, have the same length, and intersect at right angles?

The theorems you discovered in the last two exercises will be proved in Chapter 6.

1.4 Enhancing the sketch

When you worked through the example in the previous section, you were probably dissatisfied with the fact that the labels that are automatically assigned by GeoGebra do not agree with the labels we normally assign when we are doing the construction by hand. For example, when $\triangle ABC$ is a right triangle, we like the right angle to be at vertex C and we like a side of the triangle to be labeled with the lower case letter that matches the upper case letter on the opposite vertex. Your sketches will be easier for other humans to understand if you take the time to coax GeoGebra into using this familiar notation.

If you do not like the name GeoGebra assigns, you can change it. To do so, use the Move tool to select the object and then type the name you want to assign. When you type the letter or name you want to use, a box will appear with the new name in it. Click OK to confirm your choice. GeoGebra is case sensitive.

You can tidy up your diagram by hiding objects that are used as part of the construction process but are not important for the final diagram. An example of such an object is the perpendicular line in the example in §1.3. It is needed so we can be sure that the triangle is a right triangle, but once the vertices of the triangle have been determined we no longer have any need to see it. In fact, the diagram is clearer without it. The Show / Hide Object tool, which is in the toolbar below the Move Graphics View tool, can be used to hide it. First activate the tool, then click on the object you want to hide. As soon as you choose a different tool the object will disappear from the Graphics View. GeoGebra still knows it is there and it will still appear in the Algebra View, but it will not clutter up your sketch.

To control the name that is assigned to a quantity you compute, you can type an expression of the form *QuantityName* = *Formula* in the Input Bar. When you do that, the quantity will be assigned the name you give it and the name should appear in the Algebra View.

It is often helpful to display computed quantities right in the Graphics View. This can be done by use of the Insert Text tool, which is under the Slider tool in the toolbar. Type the text you want to show and then click on the object in the Algebra View that you want to display. You can either use the pull down menus in the text box or use LaTeX commands to enter a mathematical formula. If you use LaTeX, click the LaTeX formula button in the text window and enclose the LaTeX formula in $'s. GeoGebra even provides a window that contains a preview of your LaTeX formulas.

Example. Let us rework the example in the previous section. Begin by constructing two points and the segment joining them. Rename the points A and C. Construct the line that is perpendicular to $\overleftrightarrow{AC}$ at C, choose a movable point on the perpendicular line, and name it B. Hide the perpendicular line and construct the segments $\overline{AB}$ and $\overline{BC}$. Rename the segments so that $\overline{AB}$ is named c, $\overline{AC}$ is named b, and $\overline{BC}$ is named a.

In the Input Bar, type `HypotenuseSquared = c^2` and press enter; then type `SumofLegsSquared = a^2 + b^2` and press enter again. You should see both `HypotenuseSquared` and `SumofLegsSquared` in the Algebra View and they should be numerically equal (Figure 1.8). Drag A and B around and verify that the two quantities remain equal.

To display these quantities in the Graphics View, activate the Insert Text tool, click somewhere in the Graphics Window, and type `$a^2 + b^2$ =` in the text box that pops up. Then click on `SumofLegsSquared` in the Algebra View, click the button for LaTeX,

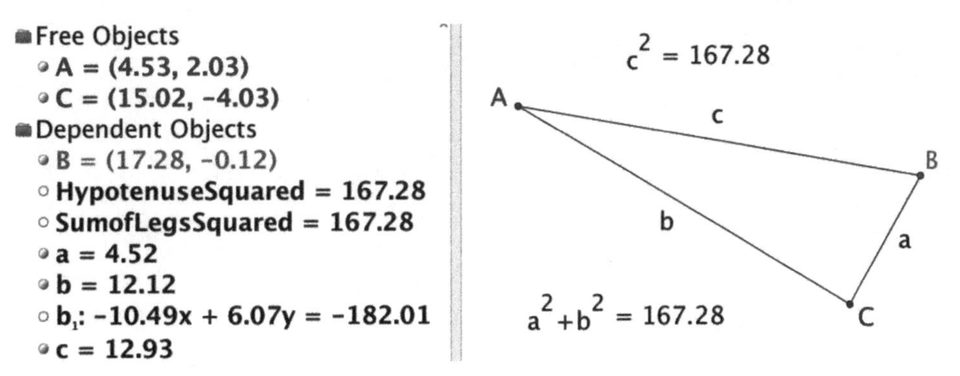

Figure 1.8. The Algebra and Geometry views of the modified example calculation

and click **OK**. You should then see the value of $a^2 + b^2$ displayed in the Graphics View, as in Figure 1.8. Display the value of c^2 in a similar way. You can now hide the Algebra View and perform the drag test entirely within the Graphics View. □

As you become a more experienced user of GeoGebra you should begin to consider subjective aspects of your sketches. If you take the time to add details that make your diagram easy to comprehend and include appropriate explanatory text, you will make your sketch more user friendly. You should strive to produce a document that the reader can understand without undue effort as well as one that is pleasing to look at.

Pay careful attention to the appearance of the objects in your diagram. Begin by hiding any unnecessary objects. To modify the appearance of remaining objects, open the **Object Properties** box, which is found in the **Edit** menu. In the left hand column you should see a list of the objects in your current sketch. Click on the name of one of the objects and then you can modify its properties by using the **Color** and **Style** tabs. Thoughtful use of colors and thicknesses to distinguish the objects in your sketches can make the diagrams easier to comprehend.

You should also add text boxes in which you give readers any information they need in order to understand your sketch. You should explain the construction, what it illustrates, and what conclusions you drew from it. In the sketches you produce as solutions to exercises, you should include text boxes in which you record observations you made or conclusions you drew from your observations.

Exercises

*1.4.1. Go back to the sketch you made for Exercise 1.3.4. Add a text box in which you explain what the diagram illustrates and what exploration reader can do to see this.

*1.4.2. Go back to the sketch you made for Exercise 1.3.7. Make use of color and line thickness to distinguish the original quadrilateral from the associated midpoint quadrilateral. Add a text box that tells readers what the sketch illustrates and what they can do with the sketch to explore this relationship.

For the remainder of the course you should try to include the kinds of enhancements discussed in this section in all of your GeoGebra sketches.

2

The Classical Triangle Centers

We begin our study of advanced Euclidean geometry by looking at several points associated with a triangle. These points are all called *triangle centers* because each of them can claim to be at the center of the triangle in a certain sense. They are *classical* in that they were known to the ancient Greeks. The classical triangle centers form a bridge between elementary and advanced Euclidean geometry. They also provide an excellent setting in which to develop proficiency with GeoGebra.

While the three classical triangle centers were known to the ancient Greeks, the ancients missed a simple relationship between them. This relationship was discovered by Leonhard Euler in the eighteenth century. Euler's theorem serves as a fitting introduction to advanced Euclidean geometry because Euler's discovery can be seen as the beginning of a revival of interest in Euclidean geometry and most of the theorems we will study in the text were discovered in the century after Euler lived. Euler's original proof of his theorem was a complicated analytic argument, but it is simple to discover (the statement of) the theorem with GeoGebra. See Chapter 7 of [6] for a nice discussion of Euler's contributions to geometry.

2.1 Concurrent lines

Definition. Three lines are *concurrent* if there is a point P such that P lies on all three of the lines. The point P is called the *point of concurrency*. Three segments are concurrent if they have an interior point in common.

Two arbitrary lines will intersect in a point—unless the lines happen to be parallel, which is unusual. Thus concurrency is an expected property of two lines. But it is rare that three lines should have a point in common. One of the surprising and beautiful aspects of advanced Euclidean geometry is the fact that many triples of lines determined by triangles are concurrent. Each of the triangle centers in this chapter is an example.

2.2 Medians and the centroid

Definition. The segment joining a vertex of a triangle to the midpoint of the opposite side is called a *median* for the triangle.

Exercises

*2.2.1. Construct a triangle $\triangle ABC$. Construct the midpoints of the sides of $\triangle ABC$. Label the midpoints D, E, and F in such a way that D lies on the side opposite A, E lies on the side opposite B, and F lies on the side opposite C. Construct the medians for $\triangle ABC$. What do the medians have in common? Use the drag test to verify that this continues to be true when the vertices of the triangle are moved around in the plane.

*2.2.2. In the preceding exercise you should have discovered that the three medians are concurrent (they have an interior point in common). Mark the point of intersection and label it G. Measure AG and GD, and then calculate AG/GD. Make an observation about the ratio. Now measure BG, GE, CG, and GF, and then calculate BG/GE and CG/GF. Leave the calculations displayed on the screen while you move the vertices of the triangle. Make an observation about the ratios.

The two exercises you just completed should have led you to discover the following theorem.

Median Concurrence Theorem. *The three medians of any triangle are concurrent; that is, if $\triangle ABC$ is a triangle and D, E, and F are the midpoints of the sides opposite A, B, and C, respectively, then $\overline{AD}$, $\overline{BE}$, and $\overline{CF}$ intersect in a common point G. Moreover, $AG = 2GD$, $BG = 2GE$, and $CG = 2GF$.*

Definition. The point of concurrency of the medians is called the *centroid* of the triangle. The centroid is usually denoted by G.

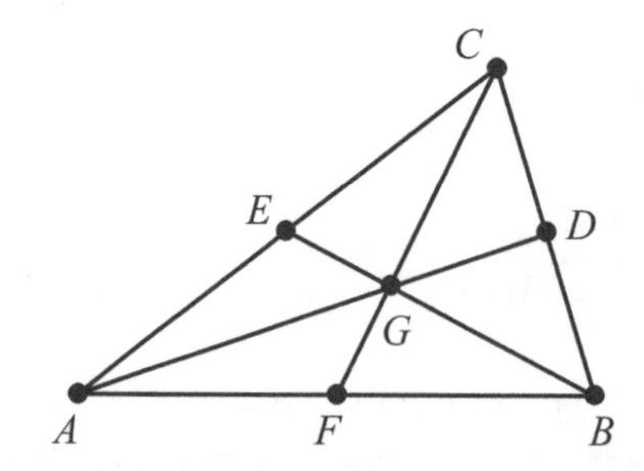

Figure 2.1. The three medians and the centroid

We will not prove the concurrency part of the theorem at this time. That proof will be postponed until after we have developed a general principle, due to Ceva, that allows us to give one unified proof of all the concurrence theorems at the same time. For now we will assume the concurrency part of the theorem and prove that the centroid divides each median in a 2:1 ratio. That proof is outlined in the next four exercises.

Exercises

*2.2.3. Construct a triangle $\triangle ABC$, then construct the midpoints of the sides $\overline{BC}$ and $\overline{AC}$ and label them as in Exercise 2.2.1. Construct the medians $\overline{AD}$ and $\overline{BE}$ and the segment $\overline{DE}$. Check, by measuring angles, that $\triangle ABC \sim \triangle EDC$. Then check that $\triangle ABG \sim \triangle DEG$.

Note. In the remaining exercises of this section we assume the notation of Exercise 2.2.1 (or 2.2.3).

2.2.4. Use theorems from Chapter 0 to prove that $\triangle ABC \sim \triangle EDC$ and that $AB = 2ED$.

2.2.5. Use theorems from Chapter 0 to prove that $\triangle ABG \sim \triangle DEG$.

2.2.6. Use the two preceding exercises to prove that $AG = 2GD$ and $BG = 2GE$. Explain how this allows you to conclude that $CG = 2GF$ as well.

*2.2.7. Each median divides the triangle into two subtriangles. For example, $\overline{AD}$ divides $\triangle ABC$ into $\triangle ADB$ and $\triangle ADC$. Verify that $\alpha(\triangle ADB) = \alpha(\triangle ADC)$. Since the two triangles share the base $\overline{AD}$ and have equal areas, they must have the same heights as well. Verify this by dropping perpendiculars from B and C to $\overleftrightarrow{AD}$, marking the feet of the perpendiculars, and then measuring the distances from B and C to the feet.

2.2.8. Let H and K denote the feet of the perpendiculars from B and C to $\overleftrightarrow{AD}$. Prove that $BH = CK$. Conclude that $\alpha(\triangle ADB) = \alpha(\triangle ADC)$.

2.2.9. Explain why the centroid is the center of mass of the triangle. In other words, explain why a triangle made of a rigid, uniformly dense material would balance at the centroid.

*2.2.10. The three medians subdivide the triangle $\triangle ABC$ into six smaller triangles (see Figure 2.1). Determine the shape of $\triangle ABC$ for which the subtriangles are congruent. Measure the areas of the subtriangles and verify that their areas are always equal (regardless of the shape of the original triangle).

2.2.11. Prove that the six subtriangles in the preceding exercise have equal areas.

2.2.12. Prove that if $\triangle ABC$ is equilateral, then the six subtriangles are all congruent to one another. Is the converse correct?

2.3 Altitudes and the orthocenter

Definition. The line determined by two vertices of a triangle is called a *sideline* for the triangle.

As we move deeper into advanced Euclidean geometry we will see that the sidelines play an increasingly important role in the geometry of a triangle.

Definition. An *altitude* for a triangle is a line through one vertex that is perpendicular to the opposite sideline of the triangle. The point at which an altitude intersects the opposite sideline of a triangle is called the *foot* of the altitude.

Exercise

*2.3.1. Construct a triangle and construct its three altitudes. Observe that no matter how the vertices of the triangle are moved around in the plane, the altitudes continue to concur.

The exercise should have confirmed the following theorem. It will be proved in Chapter 8.

Altitude Concurrence Theorem. *The three altitudes of a triangle are concurrent.*

Definition. The point of concurrency of the three altitudes is called the *orthocenter* of the triangle. It is usually denoted by H.

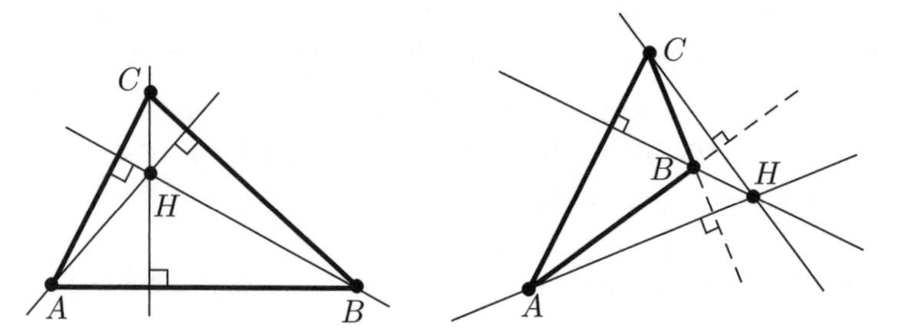

Figure 2.2. Two triangles with their altitudes and orthocenters

Exercises

*2.3.2. Construct another triangle. Mark the orthocenter of your triangle and label it H. Move one vertex and watch what happens to H. Add the centroid G to your sketch and again move the vertices of the triangle. Observe that the centroid is always located inside the triangle, but that the orthocenter can be outside the triangle or even on the triangle. (A rigorous proof that the centroid is always inside the triangle relies on technical results about betweenness that were not stated formally in Chapter 0. A proof could be based on Theorems 3.3.9 and 3.3.10 of [11], for example. Of course such a proof would be preceded by a careful definition of the interior of a triangle.)

*2.3.3. Determine by experimentation the shape of triangles for which the orthocenter is outside the triangle. Find a shape for the triangle so that the orthocenter is equal to one of its vertices. Observe what happens to the orthocenter when one vertex crosses over the sideline determined by the other two vertices. Make notes on your observations.

*2.3.4. Determine by experimentation whether or not it is possible for the centroid and the orthocenter to be the same point. If it is possible, for which triangles does this happen?

2.4 Perpendicular bisectors and the circumcenter

The perpendicular bisectors of the three sides of a triangle are concurrent. That fact will be verified experimentally in this section and will be proved in Chapter 8.

Definition. The point of concurrency of the three perpendicular bisectors of the sides of a triangle is called the *circumcenter* of the triangle.

The circumcenter is usually denoted by O. The reason for the name "circumcenter" will become clear when we study circles associated with triangles in Chapter 4.

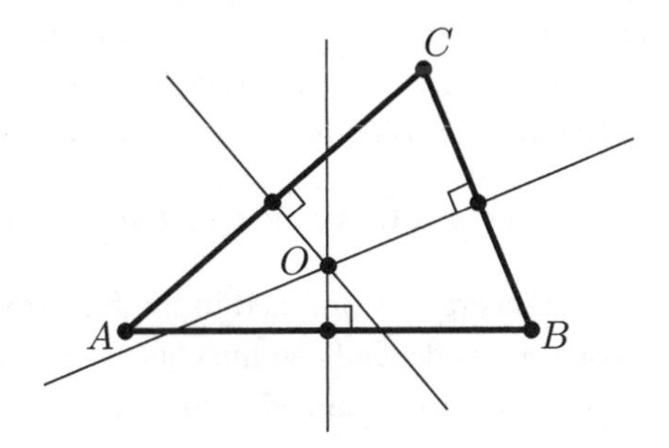

Figure 2.3. The three perpendicular bisectors and the circumcenter

Exercises

*2.4.1. Construct a triangle $\triangle ABC$ and construct the three perpendicular bisectors of its sides. These perpendicular bisectors should be concurrent and should remain concurrent when subjected to the drag test. Mark the point of concurrency and label it O.

*2.4.2. Move one vertex of the triangle and observe how the circumcenter changes. Note what happens when one vertex crosses over the line determined by the other two vertices. Find triangles that show that the circumcenter may be inside, on, or outside the triangle, depending on the shape of the triangle. Make notes on your findings.

*2.4.3. Measure the distances OA, OB, and OC and make an observation about them.

2.4.4. Prove that the circumcenter is equidistant from the vertices of the triangle.

*2.4.5. Determine by experimentation whether or not it is possible for the circumcenter and the centroid to be the same point. If it is possible, for which triangles does this happen?

2.5 The Euler line

There is a beautiful geometric relationship between the three triangle centers studied in this chapter. As mentioned at the beginning of the chapter, it was first discovered by the German mathematician Leonhard Euler (1707–1783).

Exercises

*2.5.1. Construct a triangle $\triangle ABC$ and construct the centers G, H and O. Hide any lines that were used in the construction so that only the triangle and the centers are visible. Put a line through two of them and observe that the third lies on the line. Use the drag test to verify that G, H, and O continue to be collinear when the shape of the triangle is changed.

*2.5.2. Measure the distances HG and GO. Calculate HG/GO. Leave the calculation visible on the screen as you change the shape of your triangle. Observe what happens to HG/GO as the triangle changes.

In the exercises you should have discovered the following theorem.

Euler Line Theorem. *The orthocenter H, the circumcenter O, and the centroid G of any triangle are collinear. Furthermore, G is between H and O (unless the triangle is equilateral, in which case the three points coincide) and $HG = 2GO$.*

Definition. The line through H, O, and G is called the *Euler line* of the triangle.

The proof of the Euler line theorem is outlined in the following exercises. Since the existence of the three triangle centers depends on the concurrency theorems stated earlier in the chapter, those results are implicitly assumed in the proof.

Exercises

2.5.3. Prove that a triangle is equilateral if and only if its centroid and circumcenter are the same point. If the triangle is equilateral, the centroid, the orthocenter, and the circumcenter all coincide.

2.5.4. Fill in the details in the following proof of the Euler line theorem. Let $\triangle ABC$ be a triangle with centroid G, orthocenter H, and circumcenter O. By the previous exercise, it may be assumed that $G \neq O$ (explain why). Choose a point H' on $\overrightarrow{OG}$ such that G is between O and H' and $GH' = 2OG$. The proof can be completed by showing that $H' = H$ (explain). It suffices to show that H' is on the altitude through C (explain why this is sufficient). Let F be the midpoint of $\overline{AB}$. Use Exercise 2.2.6 and the SAS Similarity Criterion to prove that $\triangle GOF \sim \triangle GH'C$. Conclude that $\overleftrightarrow{CH'} \parallel \overleftrightarrow{OF}$ and thus $\overleftrightarrow{CH'} \perp \overleftrightarrow{AB}$.

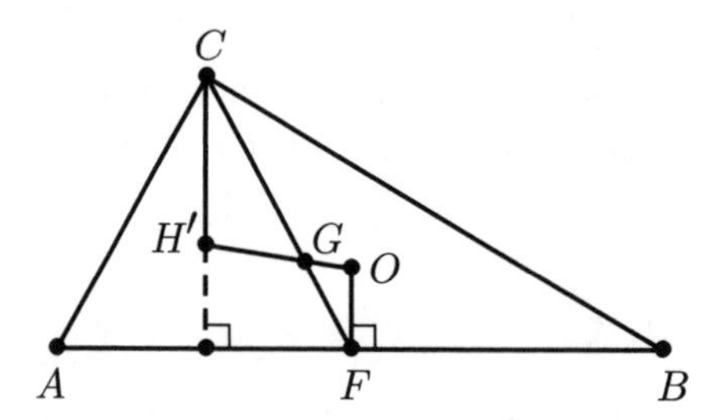

Figure 2.4. Proof of Euler line theorem

*2.5.5. Figure 2.4 shows a diagram of the proof of the Euler line theorem if the original triangle is acute. Use GeoGebra to experiment with triangles of other shapes to determine what the diagram looks like in case $\triangle ABC$ has some other shape.

2.5.6. Prove that a triangle is isosceles if and only if two medians are congruent (i.e., have the same length).

2.5.7. Prove that a triangle is isosceles if and only if two altitudes are congruent.

*2.5.8. Construct a triangle $\triangle ABC$ and the bisectors of the interior angles at A and B. Mark the points at which the angle bisectors intersect the opposite sides of the triangle and label the points D and E, respectively. The segments $\overline{AD}$ and $\overline{BE}$ are called *internal angle bisectors*. Measure the lengths of the internal angle bisectors and then measure the lengths of the sides $\overline{BC}$ and $\overline{AC}$. Use your measurements to verify the following theorem: *A triangle is isosceles if and only if two internal angle bisectors are congruent.*

The theorem in the last exercise is known as the *Steiner-Lehmus theorem* because the question was proposed by C. L. Lehmus in 1840 and the theorem was proved by the Swiss mathematician Jacob Steiner in 1842. A proof of the theorem and discussion of the interesting history of the proof may be found in §1.5 of [3].

3

Advanced Techniques in GeoGebra

This chapter introduces two features of GeoGebra that enhance its functionality. By far the most useful property of GeoGebra for our purposes is the ability to create user-defined tools, which allow routine constructions to be carried out quickly and efficiently. Check boxes make it possible to produce GeoGebra documents that illustrate a process, not just a static finished sketch.

3.1 User-defined tools

One of the most effective ways to tap the power of GeoGebra is to create your own custom-made tools to perform constructions that you will want to repeat several times. Once you learn how to make tools of your own, you should accumulate a collection of tools that you can use when you need them.

Creating a tool. Creating a new tool of your own is surprisingly easy. The first step is to start GeoGebra and perform the construction your tool will replicate. Then use the Move tool to select the objects in the sketch that you want your tool to produce as output objects. You can select multiple objects by holding down the Control key (on a PC) or the Command key (on a Macintosh). Next choose Create New Tool in the Tools menu. A dialog box will appear that contains three tabs: Output Objects, Input Objects, and Name & Icon. In the Output Objects window you should see a list of the objects you selected. You can add additional objects either by choosing them from the pulldown menu or by moving the dialog box to the side and clicking on the objects in the Graphics View. You can remove unwanted objects with the Remove tab (✘). The Input Objects window will automatically contain a list of the minimal set of objects that must be specified to produce the output objects you have chosen. You can require the user of the tool to specify more than the minimum by adding objects to the Input Objects window. Finally, click on the Name & Icon tab and enter the required information to name your tool. A generic icon will be supplied by GeoGebra. Check the Show in Toolbar box and click on Finish. Your tool has now been created and should appear in the toolbar. Following the name of the new tool GeoGebra lists the kinds of input objects that the tool requires.

Example. Open a new graphics window and construct a triangle, the midpoints of the sides, the medians, and the centroid. Perform the drag test to make sure your construction is correct. Now select the three medians and then choose Create New Tool in the Tools menu. In the dialog box that appears you should see the three medians listed as Output Objects and the three vertices of the triangle listed as Input Objects. Name the tool something like "Medians." Check the Show in Toolbar box and click on Finish. Your tool should appear in the toolbar. □

To use the tool just created, activate it in the toolbar and then click on three points in the graphics view. The result should be that three points are created where you click and the medians of the corresponding triangle are drawn. You may be surprised that the triangle itself is not shown since medians without a corresponding triangle look incomplete. In most cases, you will want the triangle to be constructed as well as the medians. You can easily create a different tool that accepts three vertices as input objects and produces the sides of triangle and the medians as output objects—go back to your original diagram and choose both the sides and the medians before you create the new tool (or add the sides in the Output Objects pane in the Create New Tool window).

Another variation that might be useful is to color and thicken the medians. If you do that before you create the tool, then medians will always have those attributes when they are constructed with your tool. This allows you to achieve a consistent appearance for medians so they will be easy to identify in your sketches.

Saving a tool. If your tools are to be useful, they must be available to you when you need them. For that reason it is important that you pay attention to how and where you save your tools. You should aim to produce a toolbox with tools that are always available to be used in your constructions.

When you create a tool as in the last example, it will be available to you during the current GeoGebra session. It is saved as part of the GeoGebra document as well and will be available the next time you open the document. But you do not want to have to open every document containing every tool you have created whenever you use GeoGebra. For that reason GeoGebra has provided a way to save tools separately.

Making the tools you create available in the toolbar during subsequent GeoGebra sessions is a two-step process. First choose Manage Tools in the Tools menu. A box will pop up that contains a list of the tools you have created. Choose the tool you want to save and click on Save As. You can then name and save the tool. The tool's file name will be given the extension .ggt to distinguish it from ordinary GeoGebra documents, which have the extension .ggb. This will save the tool as a separate document, but it still does not make it available during subsequent sessions. To accomplish that, click on Save Settings in the dialog box that appears when you choose Settings... in the Options menu. Your tool will then be available to you whenever you use GeoGebra. If you later change your mind and want to remove the tool from the toolbar, you can do that using Customize Toolbar, which is found in the Tools menu.

Exercises

*3.1.1. Create a GeoGebra tool that accepts two points as input objects and whose output object is an equilateral triangle with the given points as the endpoints of one side. Save your tool for future use.

*3.1.2. Create a GeoGebra tool that accepts two points as input objects and whose output is a square with the given points as the endpoints of one side. Save your tool for future use. [Compare Exercise 1.2.2.]

*3.1.3. Create a GeoGebra tool that accepts a segment as its input object and constructs a circle with the given segment as diameter. Save your tool for later use.

*3.1.4. Create a GeoGebra tool that constructs the centroid of a triangle. Your tool should accept three vertices of a triangle as input objects and produce the triangle and the centroid G as output objects. The tool should not display the intermediate objects (such as the medians) that were used in the construction of G.

*3.1.5. Create a GeoGebra tool that constructs the altitudes of a triangle. Your tool should accept three vertices of a triangle as input objects and produce the triangle and the altitudes as output objects. Make the tool so that the altitudes always appear in the same distinctive color.

*3.1.6. Make a GeoGebra tool that constructs the orthocenter of a triangle. Your tool should accept three vertices of a triangle as input objects and produce the triangle and the orthocenter as output objects. The tool should not display the intermediate objects that were used in the construction.

*3.1.7. Make a GeoGebra tool that constructs the circumcenter of a triangle. Your tool should accept three vertices of a triangle as input objects and produce the triangle and the circumcenter as output objects.

3.2 Check boxes

A geometric construction, whether it is done with pencil and paper or by using computer software, is a process that is not completely captured by the final sketch. You should strive to produce GeoGebra documents that reflect the dynamic nature of constructions. One thing this means is that the objects in the sketch should not all appear at once, but should become visible in the order in which they are constructed. It should be evident how each new object is related to previously constructed objects. After an object has served its purpose as a step towards the construction of some other object, you may want to hide it.

The most straightforward way to accomplish this in GeoGebra is to include "check boxes" that show or hide various objects as the boxes are checked. Check boxes can be used to show or hide either geometric objects or text boxes.

To create a check box, activate Check Box to Show / HIde Objects in the toolbar (under the Slider icon) and then click somewhere in the graphics view. A window will pop up in which you specify the objects to be controlled by the check box. You can either choose the objects from the pull down menu or you can move the dialog box to the side and click on the appropriate objects in the Graphics View. Supply a caption for the check box and then click Apply. In the Graphics Window you will see a check box with the caption you have specified. When the box is checked the objects you have specified are shown and when the box is not checked the objects are hidden. You check and uncheck it by clicking on it with the Move tool.

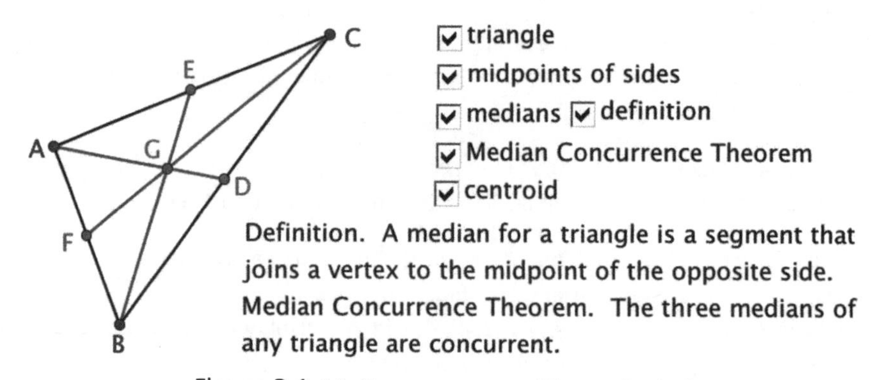

Figure 3.1. Median concurrence; all boxes checked

Example. Go back to the sketch you produced when you worked through the example in the previous section. Add check boxes that show or hide the original triangle, the midpoints of the sides, the medians, and the centroid. Then add text boxes that give the definition of median and the statement of the median concurrence theorem. Create check boxes that show hide the text boxes you just added. The resulting document should look something like that in Figure 3.1. A user encountering the document for the first time should see just the check boxes and their captions. As the check boxes are clicked, the components of the diagram, as well as the requisite definition and theorem, become visible. You might also want to add a final text box that invites the user to apply the drag test to verify median concurrence. □

Exercises

*3.2.1. Make a sketch that illustrates the construction of the orthocenter of a triangle.

*3.2.2. Make a new sketch that illustrates the Euler line theorem. First construct a triangle and then use the tools from §3.1 to create the three triangle centers. Hide any intermediate objects so that only the triangle itself and the three triangle centers are visible. Construct a line through two of the points and observe that it also passes through the third. Add hide/show check boxes for the triangle, the three centers, and the Euler line. Add text boxes and check boxes for the definitions of the three triangle centers as well as the statement of the Euler line theorem. Make sure that you have made good use of color and explanatory text so that your sketch is user friendly.

3.3 The Pythagorean theorem revisited

The Pythagorean theorem is part of elementary Euclidean geometry, but it is worth examining again because of its importance to geometry generally. In the next few exercises you will explore the statement of the theorem and then look at Euclid's proof. Even though there are hundreds of different proofs of the Pythagorean theorem, Euclid's proof remains one of the most beautiful and elegant.

Exercise

*3.3.1. Construct a right triangle. Use your tool from Exercise 3.1.2 to construct the square region on each side of the triangle. Measure the areas of the squares and verify the Pythagorean relationship.

Hint: Be careful about the order in which the input objects in your square tool are selected so that all three of the squares appear on the outside of the triangle.

The diagram of the right triangle with the three exterior squares on the sides goes by the name of *the bride's chair*. The historical reasons for the name are obscure, but it is most likely based on a mistranslation into Arabic of a Greek word from Euclid's *Elements* (see [1]).

Exercises

*3.3.2. Construct a right triangle. Use your tool from Exercise 3.1.1 to construct an equilateral triangle on each side of the right triangle. Measure the areas of the associated triangular regions. Find a relationship between the areas.

*3.3.3. Construct a right triangle. Use the tool from Exercise 3.1.3 to construct three circles whose diameters are the three sides of the triangle. (The circles will overlap.) Measure the areas of the circles and find a relationship between them. Explain how this relationship can be used to determine which is larger: a large pizza or the combination of one small and one medium pizza.

3.3.4. State a theorem that summarizes the results of the last three exercises.
Hint: Check Euclid's Proposition VI.31, if necessary.

3.3.5. Find Euclid's proof of the Pythagorean theorem on the world wide web. Make sure you understand the proof.
Hint: The Pythagorean theorem is Euclid's Proposition I.47. You can also find the proof on page 175 of [11].

*3.3.6. Create a GeoGebra sketch that illustrates Euclid's proof. The sketch should include check boxes that show objects as they are needed. It should also have text boxes that indicate how triangular regions are related. There are two kinds of relationships: the first is a shear in which one vertex of a triangle moves along a line that is parallel to the base of the triangle and the second type of movement is a rotation about a point.

The bride's chair construction can be carried out on any triangle, not just a right triangle. Specifically, if $\triangle ABC$ is a triangle, the *Vecten configuration for* $\triangle ABC$ is the union of the triangle and the outer squares built on the sides of the triangle. (See Figure 3.2.) Little is known about the French geometer Vecten after whom the configuration is named (see [1], page 4).

Exercises

*3.3.7. Create a tool that accepts a triangle as input object and returns the Vecten configuration for the triangle as output object. Be careful to make sure the squares are on the outside of the triangle.

*3.3.8. Start with the Vecten configuration for $\triangle ABC$ and mark the centers of the squares. For each vertex of $\triangle ABC$, construct the line joining it to the center of the square on the opposite side. Verify that the lines are always concurrent.

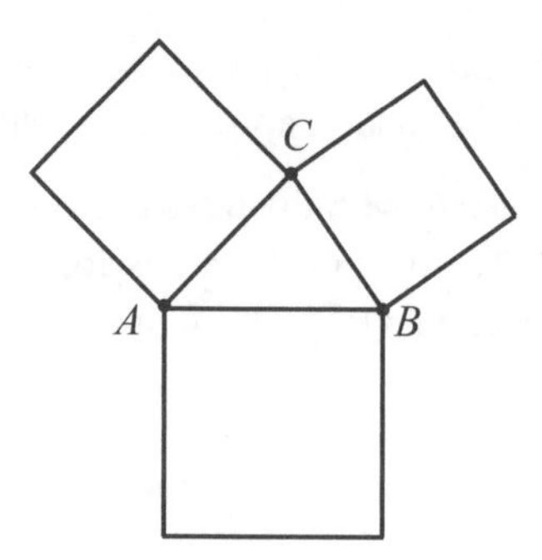

Figure 3.2. The Vecten configuration for $\triangle ABC$

Let P_A, P_B, and P_C denote the centers of the squares on the sides opposite A, B, and C, respectively. In Chapter 8 we will prove that the lines $\overleftrightarrow{AP_A}$, $\overleftrightarrow{BP_B}$, and $\overleftrightarrow{CP_C}$ are concurrent

Definition. The point of intersection of the lines $\overleftrightarrow{AP_A}$, $\overleftrightarrow{BP_B}$, and $\overleftrightarrow{CP_C}$ is called the *Vecten point* of $\triangle ABC$.

The Vecten point is a new example of a triangle center. Perhaps it would be more accurate to say that the Vecten point is an old center of a new triangle since it follows from Exercise 3.3.10 below that the Vecten point of $\triangle ABC$ is the same as the orthocenter of $\triangle P_A P_B P_C$. The construction of the Vecten point is a good example of what we have in store for us as we study advanced Euclidean geometry. Time after time we will start with a simple construction from elementary Euclidean geometry and generalize it in a natural way. When we do that we will discover unexpected and surprising relationships.

Exercises

*3.3.9. For which triangles is the Vecten point inside the triangle? For which triangles is the Vecten point on the triangle? For which triangles is the Vecten point outside the triangle?

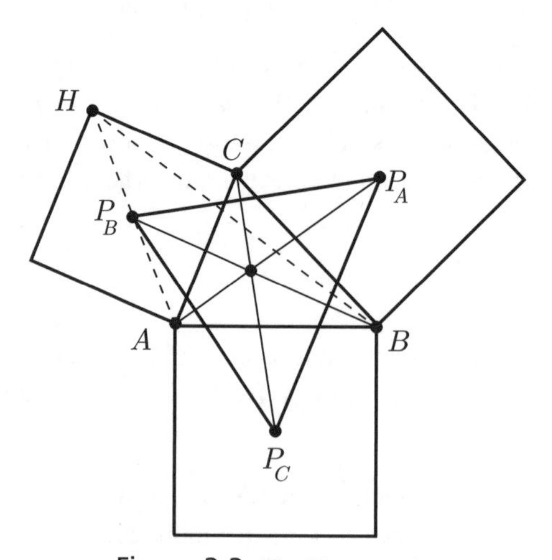

Figure 3.3. The Vecten point

3.3.10. Prove that $\overleftrightarrow{CP_C}$ is perpendicular to $\overleftrightarrow{P_AP_B}$.
Hint: Let H be the point indicated in Figure 3.3. Locate a point B' on $\overline{CB}$ such that $CB' = (1/\sqrt{2})CB$, a point H' on $\overline{CH}$ such that $CH' = (1/\sqrt{2})CH$, a point B'' on $\overline{AB}$ such that $AB'' = (1/\sqrt{2})AB$, and a point H'' on $\overline{AH}$ such that $AH'' = (1/\sqrt{2})AH$. Prove that a rotation by 45° about C transforms $\overline{H'B'}$ to $\overline{P_AP_B}$ and a rotation by 45° about A transforms $\overline{H''B''}$ to $\overline{CP_C}$.

4

Circumscribed, Inscribed, and Escribed Circles

Next we explore properties of several circles associated with a triangle. This study leads naturally to definitions of still more triangle centers. The emphasis in this chapter is again mainly GeoGebra exploration, with formal proofs of most results postponed until later. The chapter ends with a proof of Heron's formula for the area of a triangle.

4.1 The circumscribed circle and the circumcenter

The first circle we study is the circumscribed circle. It can be thought of as the smallest circle that contains a given triangle.

Definition. A circle that contains all three vertices of the triangle $\triangle ABC$ is said to *circumscribe* the triangle. The circle is called the *circumscribed circle* or simply the *circumcircle* of the triangle. The radius of the circumscribed circle is called the *circumradius*.

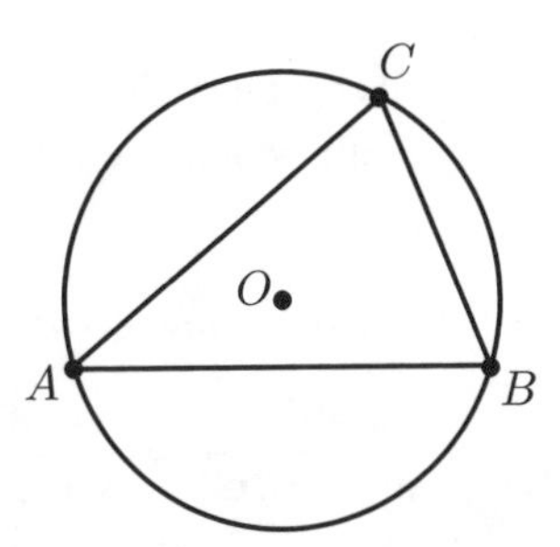

Figure 4.1. The circumcenter and the circumcircle

It is, of course, no accident that the term circumcenter was used as the name of one of the triangle centers that was introduced in Chapter 2. In fact, Exercise 2.4.4 shows that the vertices of a triangle are equidistant from the circumcenter of the triangle, so they all lie on a circle centered at the circumcenter. It follows that every triangle has a circumcircle and the circumcenter is its center. These observations are summarized in the following theorem.

Theorem. *Every triangle has a unique circumscribed circle. The circumcenter is the center of the circumscribed circle.*

It should be noted that the definition of triangle includes the assumption that the vertices are noncollinear. Thus the real assertion is that three noncollinear points determine a circle. Furthermore, the fact that we know how to locate the center of the circumcircle means we can construct the circumscribed circle itself.

Exercises

*4.1.1. Make a tool that constructs the circumcircle of a triangle. The tool should accept three noncollinear points as input objects and produce a circle containing the three points as output object. Apply the drag test to make sure your tool is robust enough so that the circle you construct continues to go through the vertices of the triangle even when they are moved.

*4.1.2. Use the tool you made in the previous exercise to explore the circumscribed circle of various triangles.

(a) For which triangles is the circumcenter inside the triangle?

(b) For which triangles is the circumcenter on the edge of the triangle?

(c) For which triangles is the circumcenter ouside the triangle?

(d) What happens to the circumcircle when one vertex of the triangle is moved across the sideline determined by the other two vertices?

Make notes on your observations.

4.1.3. Prove that the circumcircle is unique. That is, prove that there can be at most one circle that passes through three noncollinear points. How many circles pass through two given points?
Hint: Begin by proving that a circle that passes through A, B, and C must have the circumcenter as its center.

*4.1.4. Construct a triangle $\triangle ABC$, its circumcircle γ, and measure the circumradius R of $\triangle ABC$. Verify the following result, which extends the usual Law of Sines that you learned in high school.

Extended Law of Sines. *If $\triangle ABC$ is a triangle with circumradius R, then*

$$\frac{BC}{\sin(\angle BAC)} = \frac{AC}{\sin(\angle ABC)} = \frac{AB}{\sin(\angle ACB)} = 2R.$$

Hint: You can calculate $\sin(\angle BAC)$, etc., in the algebra window as well as the required quotients and then display the results of those calculations, along with appropriate text, in the graphics window.

4.1.5. Prove the Extended Law of Sines.
Hint: Let γ be the circumscribed circle of $\triangle ABC$ and let D be the point on γ such that $\overline{DB}$ is a diameter of γ. Prove that $\angle BAC \cong \angle BDC$. Use that result to prove that $\sin(\angle BAC) = BC/2R$. The other proofs are similar.

4.2 The inscribed circle and the incenter

The second circle we study is called the *inscribed circle*, or simply the *incircle*. It is the opposite of the circumcircle in the sense that it is the largest circle that is contained in the triangle.

Exercises

*4.2.1. Construct a triangle and the bisectors of its interior angles. Verify that the three angle bisectors are always concurrent. The point of concurrency is called the *incenter* of the triangle. Experiment with triangles of different shapes to determine whether the incenter can ever be on the triangle or outside the triangle.

*4.2.2. Construct another triangle and its incenter. Label the incenter I; keep it visible but hide the angle bisectors. Experiment with the triangle and the incenter to answer the following questions.

(a) Are there triangles for which the incenter equals the circumcenter? What shape are they?

(b) Are there triangles for which the incenter equals the centroid? What shape are they?

*4.2.3. Construct another triangle and its incenter. Label the incenter I; keep it visible but hide the angle bisectors. For each side of the triangle, construct a line that passes through the incenter and is perpendicular to the sideline. Mark the feet of the perpendiculars and label them X, Y and Z as indicated in Figure 4.2. Measure the distances IX, IY, and IZ and observe that they are equal. This number is called the *inradius* of the triangle.

*4.2.4. Hide the perpendiculars in your sketch from the last exercise, but keep the points I, X, Y, and Z visible. Construct the circle with center I and radius equal to the inradius. Observe that this circle is tangent to each of the sides of the triangle. This circle is the inscribed circle for the triangle.

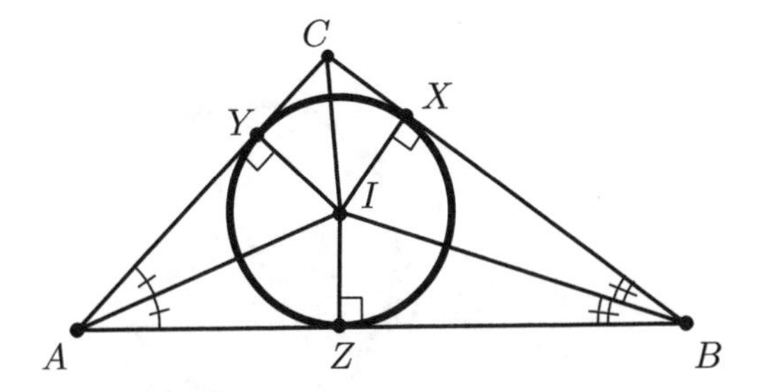

Figure 4.2. The incenter and the incircle

We now state a theorem and definition that formalize some of the observations you made in the exercises above.

Angle Bisector Concurrence Theorem. *If $\triangle ABC$ is any triangle, the bisectors of the interior angles of $\triangle ABC$ are concurrent. The point of concurrency is equidistant from the sides of the triangle.*

Definition. The point of concurrency of the bisectors is called the *incenter* of the triangle. The distance from the incenter to the sides of the triangle is the *inradius*. The circle that has its center at the incenter and is tangent to each of the sides of the triangle is called the *inscribed circle* or the *incircle* of the triangle.

The concurrency part of the theorem will be assumed for now, but it will be proved in Chapter 8. You will prove the rest of the theorem in the next exercises. Since each angle bisector is in the interior of its angle, the point at which the three angle bisectors concur is in the interior of all three angles and is therefore always in the interior of the triangle.

Exercises

4.2.5. Prove that the point at which any two interior angle bisectors intersect is equidistant from all three sidelines of the triangle.

*4.2.6. Create a tool that constructs the incenter of a triangle.

*4.2.7. Create a tool that constructs the inscribed circle of a triangle.

4.3 The escribed circles and the excenters

The incircle is not the only circle that is tangent to all three sidelines of the triangle. A circle that is tangent to all three sidelines of a triangle is called an *equicircle* (or *tritangent circle*) for the triangle. In this section we will construct and study the remaining equicircles.

Definition. A circle that is outside the triangle and is tangent to all three sidelines of the triangle is called an *escribed circle* or an *excircle*. The center of an excircle is called an *excenter* for the triangle.

There are three excircles, one opposite each vertex of the triangle. The excircle opposite vertex A is shown in Figure 4.3; it is called the A-*excircle* and is usually denoted γ_A. There is also a B-excircle γ_B and a C-excircle γ_C.

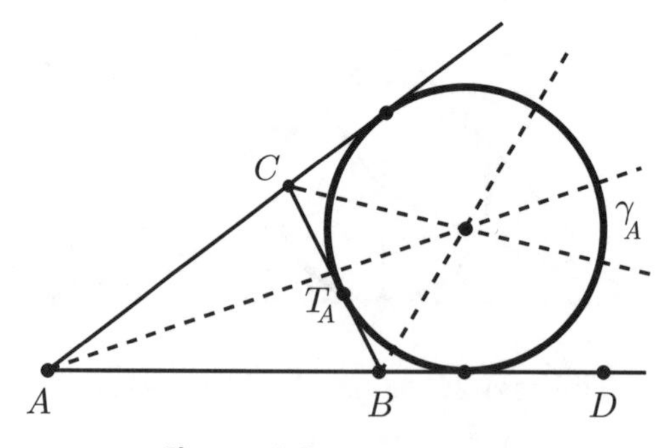

Figure 4.3. The A-excircle

Figure 4.3 indicates how the A-excircle is constructed. The A-excenter is the point at which the bisector of the interior angle at A and the bisectors of the exterior angles at B and C concur. (There are two exterior angles at B; when constructing the A-excenter use the exterior angle formed by extending $\overline{AB}$. At vertex C use the exterior angle formed by extending $\overline{AC}$.) Once the excenter has been located, a radius of the excircle is obtained by dropping a perpendicular from the excenter to one of the sidelines and marking the foot of the perpendicular.

Exercises

*4.3.1. Construct a triangle $\triangle ABC$, the rays $\overrightarrow{AC}$ and $\overrightarrow{AB}$, and the three angle bisectors shown in Figure 4.3. Experiment with triangles of different shapes to verify that the angle bisectors are always concurrent.

*4.3.2. Create a tool that constructs an excircle. Your tool should accept the three vertices of the triangle as input objects and should construct the excircle opposite the first vertex selected as its output object. Be careful to make sure that the tool always constructs the correct equicircle.

*4.3.3. Construct a triangle, its three sidelines, and its three excircles. Explore triangles of different shapes to learn what the configuration of excircles looks like in different cases. What happens to the excircles when one vertex passes over the sideline determined by the other two vertices?

The fact that the three angle bisectors are always concurrent will be proved in Chapter 8. Assuming that result, the following exercise shows that the point of concurrency is an excenter.

Exercise

4.3.4. Prove that the point at which any two of the angle bisectors shown in Figure 4.3 intersect is equidistant from all three sidelines of the triangle.

4.4 The Gergonne point and the Nagel point

There are two additional triangle centers that are associated with the inscribed and escribed circles.

Exercises

*4.4.1. Construct a triangle $\triangle ABC$ and its incircle. Mark the three points at which the incircle is tangent to the triangle. The point of tangency opposite to vertex A should be labeled X, the point of tangency opposite B should be labeled Y, and the one opposite C should be labeled Z. Construct the segments $\overline{AX}$, $\overline{BY}$, and $\overline{CZ}$. Observe that they are concurrent, regardless of the shape of the triangle. The point of concurrency is called the *Gergonne point* of the triangle. It is denoted Ge. (Don't confuse it with the centroid G.)

*4.4.2. Construct a triangle, the incenter I, and the Gergonne point Ge. For which triangles is $I = Ge$?

*4.4.3. Construct a triangle and its three excircles. Mark the three points at which the excircles are tangent to the triangle. Use the label T_A for the point at which the A-excircle is tangent to $\overline{BC}$, T_B for the point at which the B-excircle is tangent to $\overline{AC}$, and T_C for the point at which the C-excircle is tangent to $\overline{AB}$. (Type `T_A` in the GeoGebra text box to get the label T_A.) Construct the segments $\overline{AT_A}$, $\overline{BT_B}$, and $\overline{CT_C}$. Observe that the three segments are always concurrent, regardless of the shape of the triangle. The point of concurrency is call the *Nagel point* of the triangle. It is denoted Na.

*4.4.4. Is the Nagel point ever equal to the Gergonne point? If so, for which triangles?

*4.4.5. Which of the three points I, *Ge*, or *Na* lies on the Euler line? Do the points you have identified lie on the Euler line for every triangle, or only for some triangles?

The Nagel point is named for the German high school teacher and geometer Christian Heinrich von Nagel (1803–1882) while the Gergonne point is named for the French mathematician Joseph Diaz Gergonne (1771–1859). Each of the Nagel and Gergonne points is defined as the point at which three segments are concurrent. You have accumulated GeoGebra evidence that these segments concur, but we have not actually proved that. The two concurrence theorems that allow us to rigorously define the Gergonne and Nagel points will be proved in Chapter 8. In that chapter we will also show that the two points are "isotomic conjugates" of one another (definitions later).

4.5 Heron's formula

In this section we use the geometry of the incircle and the excircle to derive a famous formula that expresses the area of a triangle in terms of the lengths of its sides. This formula is named for the ancient Greek geometer Heron of Alexandria who lived from approximately AD 10 until about AD 75. Even though the formula is commonly attributed to Heron, it was probably already known to Archimedes (287–212 BC).

Let us begin with some notation that will be assumed for the remainder of this section. Fix $\triangle ABC$. Let I be the center of the incircle and let E be the center of the A-excircle. The feet of the perpendiculars from I and E to the sidelines of $\triangle ABC$ are labeled X, Y, Z, T, U, and V, as indicated in Figure 4.4.

Define $a = BC, b = AC$, and $c = AB$. We will use r to denote the inradius of $\triangle ABC$ and r_a to denote the radius of the A-excircle.

Definition. The *semiperimeter* of $\triangle ABC$ is $s = (1/2)(a + b + c)$.

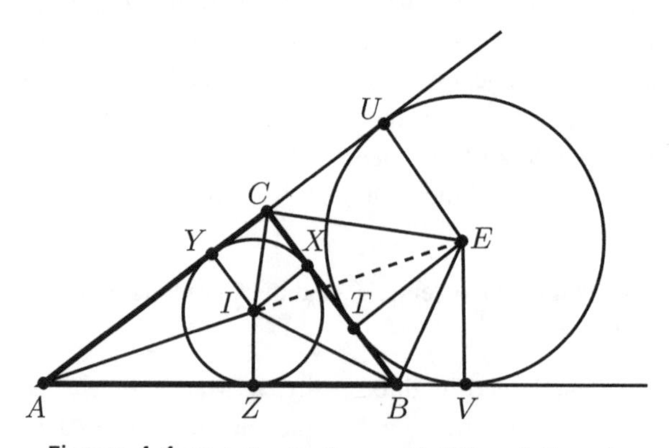

Figure 4.4. Notation for the proof of Heron's formula

Exercises

4.5.1. Prove that the area of $\triangle ABC$ satisfies $\alpha(\triangle ABC) = sr$.

4.5.2. By the external tangents theorem, $AZ = AY$, $BZ = BX$, $CX = CY$, $BV = BT$, and $CT = CU$. To simplify notation, let $z = AZ$, $x = BX$, $y = CY$, $u = CU$, and $v = BV$.

(a) Prove that $x + y + z = s$, $x + y = a$, $y + z = b$, and $x + z = c$.

(b) Prove that $x = s - b$, $y = s - c$, and $z = s - a$.

(c) Prove that $u + v = x + y$ and $x + z + v = y + z + u$.

(d) Prove that $u = x$ and $v = y$.

4.5.3. Prove that $AV = s$.

4.5.4. Use similar triangles to prove that $r_a/s = r/(s - a)$.

4.5.5. Prove that $\mu(\angle ZBI) + \mu(\angle EBV) = 90°$. Conclude that $\triangle ZBI \sim \triangle VEB$.

4.5.6. Prove that $(s - b)/r = r_a/(s - c)$.

4.5.7. Combine Exercises 4.5.1, 4.5.4, and 4.5.6 to prove Heron's Formula,

$$\alpha(\triangle ABC) = \sqrt{s(s-a)(s-b)(s-c)}.$$

5

The Medial and Orthic Triangles

We now investigate constructions of new triangles from old. We begin by studying two specific examples of such triangles, the medial triangle and the orthic triangle, and then generalize the constructions in two different ways.

Note on terminology. A median of a triangle was defined to be the segment from a vertex of the triangle to the midpoint of the opposite side. While this is usually the appropriate definition, there are occasions in this chapter when it is more convenient to define the median to be the line determined by the vertex and midpoint rather than the segment joining them. The reader should use whichever definition fits the context. In the same way an altitude of a triangle is usually to be interpreted as a line, but may occasionally be thought of as the segment joining a vertex to a point on the opposite sideline.

5.1 The medial triangle

Throughout this section $\triangle ABC$ is a triangle, D is the midpoint of $\overline{BC}$, E is the midpoint of the segment $\overline{AC}$, and F is the midpoint of $\overline{AB}$ (Figure 5.1).

Definition. The triangle $\triangle DEF$ is the *medial triangle* of $\triangle ABC$.

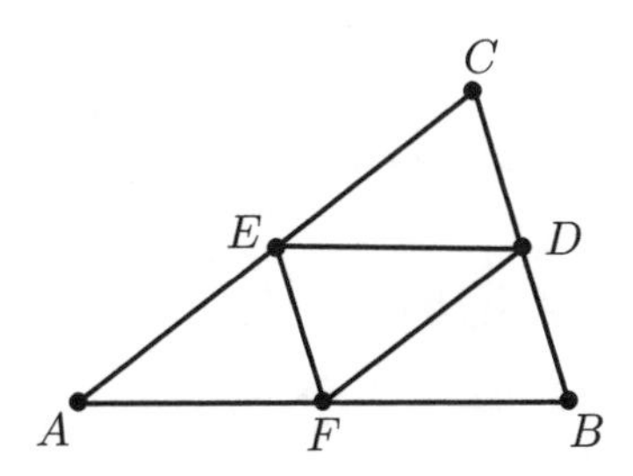

Figure 5.1. The medial triangle

Exercises

*5.1.1. Create a tool that constructs the medial triangle of a given triangle.

5.1.2. Prove that the medial triangle divides the original triangle into four congruent triangles.

5.1.3. Prove that the original triangle $\triangle ABC$ and medial triangle $\triangle DEF$ have the same medians.

5.1.4. Prove that the perpendicular bisectors of the sides of the original triangle $\triangle ABC$ are the same as the altitudes of the medial triangle $\triangle DEF$.

The last exercise shows that the circumcenter of $\triangle ABC$ is equal to the orthocenter of the medial triangle $\triangle DEF$.

Construction. Start with a triangle $\triangle ABC$ and construct a line through each vertex that is parallel to the opposite side of the triangle. Let A'', B'', and C'' be the points at which these three lines intersect. (See Figure 5.2.) The triangle $\triangle A''B''C''$ is called the *anticomplementary triangle* of $\triangle ABC$. (The term *antimedial* would be more appropriate since the construction is the opposite of the construction of the medial triangle. But the term anticomplementary is firmly established in the literature.)

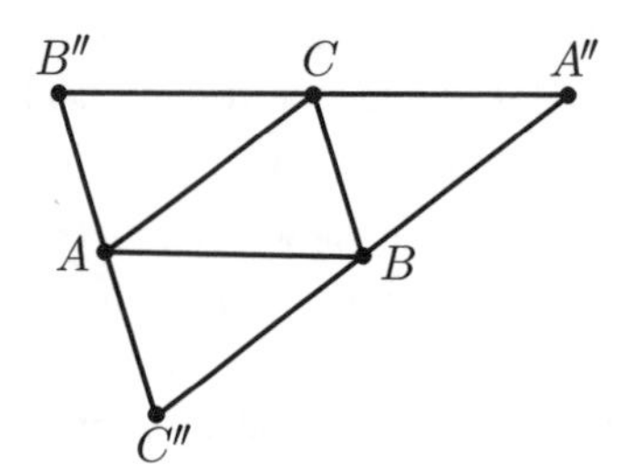

Figure 5.2. $\triangle A''B''C''$ is the anticomplementary triangle of $\triangle ABC$

Exercises

5.1.5. Prove that $\triangle ABC$ is the medial triangle of $\triangle A''B''C''$.

*5.1.6. Create a tool that constructs the anticomplementary triangle of a given triangle.

*5.1.7. Let L be the orthocenter of the anticomplementary triangle $\triangle A''B''C''$. Use GeoGebra to verify that L lies on the Euler line of the original triangle $\triangle ABC$. The point L is another triangle center for $\triangle ABC$; it is known as the *de Longchamps point* of $\triangle ABC$ for G. de Longchamps (1842–1906).

*5.1.8. Let H be the orthocenter, O the circumcenter, and L the de Longchamps point of $\triangle ABC$. Verify that O is the midpoint of the segment $\overline{HL}$.

5.2 The orthic triangle

Let $\triangle ABC$ be a triangle. We will use A' to denote the foot of the altitude through A, B' to denote the foot of the altitude through B, and C' to denote the foot of the altitude through C. Note that A' is a point on the sideline $\overleftrightarrow{BC}$ and is not necessarily on the side $\overline{BC}$. Similar comments apply to B' and C'.

Definition. The triangle $\triangle A'B'C'$ is the *orthic triangle* of $\triangle ABC$.

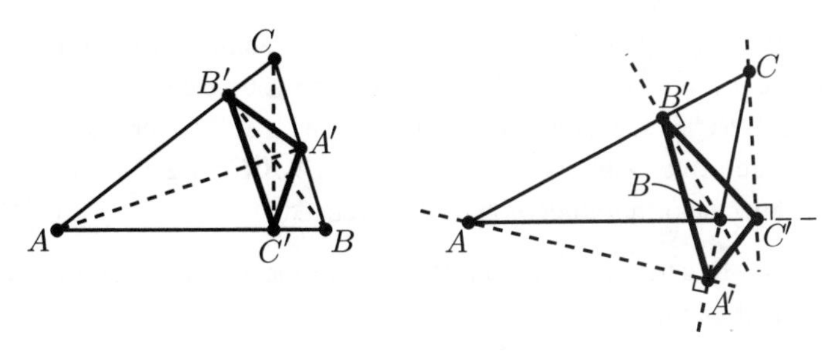

Figure 5.3. Two orthic triangles

Exercises

*5.2.1. Create a tool that constructs the orthic triangle of a triangle.

*5.2.2. Construct a triangle $\triangle ABC$ and use the tool you just created to construct the orthic triangle $\triangle A'B'C'$. Move the vertices of $\triangle ABC$ and observe what happens to the orthic triangle.

(a) Under what conditions is $\triangle A'B'C'$ contained in $\triangle ABC$?

(b) Is it possible for the orthic triangle to be completely outside the original triangle?

(c) The orthic "triangle" may not be a triangle at all since its vertices can be collinear. Under what conditions does the orthic triangle degenerate into a line segment?

(d) Is it possible for exactly one of the vertices of the orthic triangle to equal one of the vertices of the original triangle?

Make notes on your observations.

5.2.3. Let $\triangle ABC$ be a triangle.

(a) Prove that B' lies in the interior of $\overline{AC}$ if and only if $\angle BAC$ and $\angle ACB$ are both acute.

(b) Prove that every triangle must have at least one altitude with its foot on the triangle.

(c) Prove that if $\angle ABC$ is obtuse, then A' and C' lie outside $\triangle ABC$.

*5.2.4. Investigate the relationship between the measures of the angles $\angle ABC$, $\angle AB'C'$, and $\angle A'B'C$. Find two other triples of angles in the diagram whose measures are related in the same way.

5.2.5. Prove that sides of the orthic triangle cut off three triangles from $\triangle ABC$ that are all similar to $\triangle ABC$. Specifically, if $\triangle A'B'C'$ is the orthic triangle for $\triangle ABC$ and $\triangle ABC$ is not a right triangle, then

$$\triangle ABC \sim \triangle AB'C' \sim \triangle A'BC' \sim \triangle A'B'C.$$

Hint: First prove that $\triangle AB'B \sim \triangle AC'C$ and then apply the SAS similarity criterion to prove $\triangle ABC \sim \triangle AB'C'$. Use similar proofs to show that $\triangle ABC \sim \triangle A'BC'$ and $\triangle ABC \sim \triangle A'B'C$.

*5.2.6. Construct a triangle $\triangle ABC$, the three altitudes, and the orthic triangle $\triangle A'B'C'$. Now add the incenter and incircle of $\triangle A'B'C'$ to your sketch. Verify that the orthocenter of $\triangle ABC$ is the same as the incenter of $\triangle A'B'C'$ provided all the angles in $\triangle ABC$ are acute. What is the incenter of $\triangle A'B'C'$ if $\angle BAC$ is obtuse? Try to determine how the orthocenter of $\triangle ABC$ is related to $\triangle A'B'C'$ if $\angle BAC$ is obtuse.

5.2.7. Let $\triangle ABC$ be a triangle in which all three interior angles are acute and let $\triangle A'B'C'$ be the orthic triangle.

(a) Prove that the altitudes of $\triangle ABC$ are the angle bisectors of $\triangle A'B'C'$.

(b) Prove that the orthocenter of $\triangle ABC$ is the incenter of $\triangle A'B'C'$.

(c) Prove that A is the A'-excenter of $\triangle A'B'C'$.

Hint: Use Exercise 5.2.5.

5.2.8. Let $\triangle ABC$ be a triangle such that $\angle BAC$ is obtuse and let $\triangle A'B'C'$ be the orthic triangle.

(a) Prove that A is the incenter of $\triangle A'B'C'$.

(b) Prove that the orthocenter of $\triangle ABC$ is the A'-excenter of $\triangle A'B'C'$.

5.3 Cevian triangles

The construction of the medial triangle and the orthic triangle are both examples of a more general construction. Start with a triangle $\triangle ABC$ and a point P that is not on any of the sidelines of $\triangle ABC$. Let L be the point at which $\overleftrightarrow{AP}$ intersects $\overleftrightarrow{BC}$, let M be the point at which $\overleftrightarrow{BP}$ intersects $\overleftrightarrow{AC}$, and let N be the point at which $\overleftrightarrow{CP}$ intersects $\overleftrightarrow{AB}$.

Definition. The triangle $\triangle LMN$ is the *Cevian triangle* for $\triangle ABC$ associated with the point P.

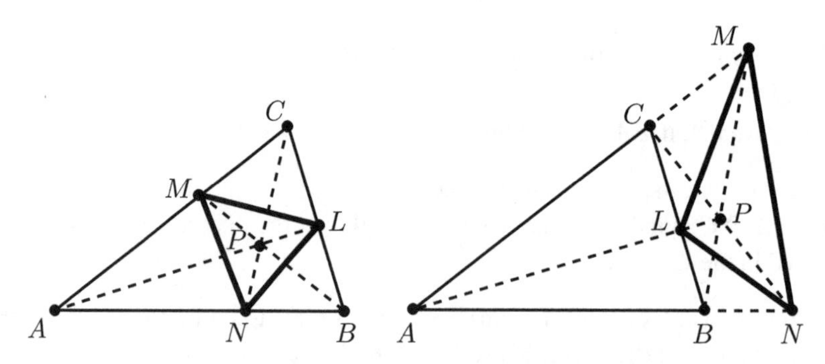

Figure 5.4. Two Cevian triangles associated with different points P

The medial triangle is the special case of a Cevian triangle in which P is the centroid of $\triangle ABC$ and the orthic triangle is the special case in which P is the orthocenter of $\triangle ABC$. Cevian triangles are named for Giovanni Ceva, who studied when three lines through the vertices of a triangle are concurrent. Because of Ceva's work, such lines are called *Cevian lines* (or simply *cevians*) for the triangle. We will investigate Ceva's theorem in Chapter 8.

Exercises

*5.3.1. Create a tool that constructs the Cevian triangle associated with a triangle $\triangle ABC$ and point P.

5.3.2. What happens if you attempt to construct a Cevian triangle associated with a point P that is on a sideline?

*5.3.3. **Preview of Ceva's theorem.** Construct a triangle $\triangle ABC$ and the Cevian triangle associated with the point P. Compute the quantity

$$d = \frac{AN}{NB} \cdot \frac{BL}{LC} \cdot \frac{CM}{MA}.$$

Observe that d has the same value regardless of the positions of A, B, C, and P (provided that none of the distances in the expression is 0).

*5.3.4. **Preview of Menelaus's theorem.** Construct a triangle $\triangle ABC$ and the Cevian triangle associated with the point P. Let A'' be the point of intersection of the lines $\overleftrightarrow{BC}$ and $\overleftrightarrow{MN}$, let B'' be the point of intersection of $\overleftrightarrow{AC}$ and $\overleftrightarrow{LN}$, and let C'' be the point of intersection of the lines $\overleftrightarrow{AB}$ and $\overleftrightarrow{LM}$. (These points may not exist if any two of the lines are parallel, but you can always move the vertices of $\triangle ABC$ slightly so that the lines do intersect.) Verify that A'', B'', and C'' are collinear.

5.4 Pedal triangles

There is a second way in which to generalize the construction of the medial and orthic triangles that leads to another useful class of triangles. Start with a triangle $\triangle ABC$ and point P that is not on any of the sidelines of the triangles. Drop perpendiculars from P to each of the sidelines of the triangle. Let A' be the foot of the perpendicular to the sideline opposite A, let B' be the foot of the perpendicular to the sideline opposite B, and let C' be the foot of the perpendicular to the sideline opposite C.

Definition. The triangle $\triangle A'B'C'$ is the *pedal triangle* associated with $\triangle ABC$ and P.

The orthic triangle is the pedal triangle associated with the orthocenter. Since the midpoints of the sides of the triangle are the feet of the perpendiculars from the circumcenter, we see that the medial triangle is also a pedal triangle; it is the pedal triangle associated with the circumcenter.

Exercises

*5.4.1. Create a tool that constructs the pedal triangle associated with a triangle $\triangle ABC$ and point P.

*5.4.2. Use your incenter tool and the pedal triangle tool to construct the pedal triangle associated with the incenter of the triangle $\triangle ABC$. Verify that the inscribed circle for the original triangle is the same as the circumscribed circle for the pedal triangle.

*5.4.3. Construct a triangle and choose a point P in the interior of the triangle. Use your pedal triangle tool to construct the pedal triangle with respect to P. Now construct

the pedal triangle of the pedal triangle with respect to the same point P. This new triangle is called the *second pedal* triangle. There is also a *third pedal triangle*, which is the pedal triangle of the second pedal triangle (with respect to the same point P). Verify that the third pedal triangle is similar to the original triangle.
Hint: This construction will work best if your pedal triangle returns the vertices of the pedal triangle as output objects, not just the triangle itself.

*5.4.4. **Preview of Simson's theorem.** Construct a triangle $\triangle ABC$, the circumscribed circle γ for $\triangle ABC$, and a point P that is not on any of the sidelines of the triangle. Verify that the vertices of the pedal triangle are collinear if and only if P lies on γ.

5.4.5. Let $\triangle ABC$ be a triangle, let P be a point, and let A', B', and C' be the feet of the perpendiculars from P to the sidelines opposite A, B, and C, respectively. Prove that

$$B'C' = \frac{BC \cdot AP}{2R}, \ A'C' = \frac{AC \cdot BP}{2R}, \ \text{and } A'B' = \frac{AB \cdot CP}{2R},$$

where R is the circumradius of $\triangle ABC$.
Hint: Use the converse to Thales' theorem to prove that B' and C' lie on the circle with diameter $\overline{AP}$. Conclude that P is on the circumcircle of $\triangle AB'C'$. Apply the Extended Law of Sines (page 40) to the triangles $\triangle ABC$ and $\triangle AB'C'$ to get $BC/\sin(\angle BAC) = 2R$ and $B'C'/\sin(\angle BAC) = AP$. Solve for $B'C'$ to obtain the first equation. The other two equations are proved similarly.

The result in Exercise 5.4.3 was discovered by J. Neuberg in 1892. The line containing the three collinear feet of the perpendiculars in Exercise 5.4.4 is called a *Simson line* for the triangle. It is named for Robert Simson (1687–1768). We will prove Simson's theorem in Chapter 11. Exercise 5.4.5 is a technical fact that will be needed in the proof of Ptolemy's theorem in that same chapter.

6

Quadrilaterals

The geometric objects we have studied until now have been relatively simple: just lines, triangles, and circles. In the remainder of the book we will need to use polygons with more sides; specifically, we will study four-sided and six-sided figures. This chapter contains the necessary information about four-sided polygons.

6.1 Basic definitions

Let us begin by repeating the definitions from Chapter 0. Four points A, B, C, and D such that no three are collinear determine a *quadrilateral* denoted by $\square ABCD$. It is defined to be the union of four segments:

$$\square ABCD = \overline{AB} \cup \overline{BC} \cup \overline{CD} \cup \overline{DA}.$$

The four segments are the *sides* of the quadrilateral and the points A, B, C, and D are the *vertices* of the quadrilateral. The sides $\overline{AB}$ and $\overline{CD}$ are called *opposite sides* of the quadrilateral as are the sides $\overline{BC}$ and $\overline{AD}$. Two quadrilaterals are *congruent* if there is a correspondence between their vertices so that all four corresponding sides are congruent and all four corresponding angles are congruent.

In the past, the term *quadrangle* was frequently used for what we now call a quadrilateral. The difference between the two terms is that one emphasizes the fact that the figure has four sides while the other emphasizes the fact that it contains four angles. In the same way either of the terms *triangle* or *trilateral* can be used to name a three-sided or three-angled figure. We will follow current practice in using the terms triangle and quadrilateral even though it could be argued that this is inconsistent terminology. (See [3, page 52].) The duality between sides and vertices will be discussed again in Chapter 9.

A quadrilateral separates the plane into an inside and an outside. A first course on the foundations of geometry will often avoid defining the interior of a general quadrilateral because defining it precisely requires care. It is customary in such a course to define the interior of a quadrilateral only if the quadrilateral is convex. A rigorous definition of interior for general quadrilaterals belongs to the branch of mathematics called *topology* since it uses the famous Jordan curve theorem, which asserts that a closed curve separates the plane into an inside and an outside. In this course we will not worry about such foundational issues, but will accept the intuitively obvious fact that a quadrilateral in the plane has an interior. GeoGebra has no qualms about this either and the **Polygon** tool assigns an interior to a general quadrilateral.

6.2 Convex and crossed quadrilaterals

We will distinguish three classes of quadrilaterals: convex, concave, and crossed. There are several equivalent ways to define the most important of these, the convex quadrilaterals.

A quadrilateral is *convex* if each vertex lies in the interior of the opposite angle. Specifically, the quadrilateral $\square ABCD$ is convex if A is in the interior of $\angle BCD$, B is in the interior of $\angle CDA$, C is in the interior of $\angle DAB$, and D is in the interior of $\angle ABC$. This is significant because it allows us to use the additivity of angle measure. For example, if $\square ABCD$ is convex, then $\mu(\angle ABC) = \mu(\angle ABD) + \mu(\angle DBC)$. Equivalently, a quadrilateral is convex if the region associated with the quadrilateral is convex in the sense that the line segment joining any two points in the region is completely contained in the region. It can also be shown that a quadrilateral is convex if and only if the two diagonals intersect at a point that is in the interior of both diagonals [11, Theorem 4.6.9].

Our definition of quadrilateral says nothing about how the interiors of the sides intersect. The requirement that no three vertices are collinear prevents adjacent sides from intersecting at any point other than their common endpoint, but opposite sides can intersect. We will call a quadrilateral whose sides intersect at an interior point a *crossed* quadrilateral. It is easy to see that a convex quadrilateral cannot be crossed, so we have defined two disjoint collections of quadrilaterals. Any quadrilateral that is neither convex nor crossed will be called a *concave* quadrilateral.

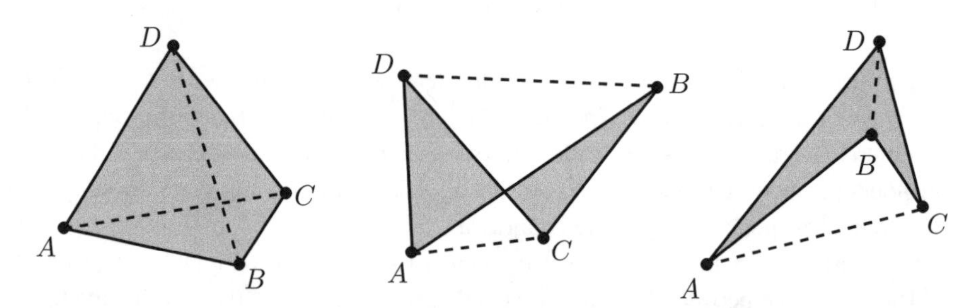

Figure 6.1. Convex, crossed, and concave

Figure 6.1 shows three quadrilaterals $\square ABCD$; the first is convex, the second is crossed, and the third is concave. In each the interior is shaded and the two diagonals are shown as dashed lines. The three kinds of quadrilaterals can be distinguished by their diagonals. The diagonals of a convex quadrilateral are both inside the quadrilateral, the diagonals of a crossed quadrilateral are both outside the quadrilateral, while one diagonal of a concave quadrilateral is inside and the other is outside.

We assume that every trapezoid is convex [11, Exercise 4.6.4]. In particular, every parallelogram is convex.

Exercises

*6.2.1. Use GeoGebra's **Polygon** tool to construct three quadrilaterals, one of each kind, and their interiors. Observe that both the quadrilateral and the interior depend on the order in which the vertices are specified.

6.2.2. Review Exercise 1.3.7. Prove the following theorem.

Varignon's Theorem. *The midpoint quadrilateral of any quadrilateral is a parallelogram.*

Does your proof work for concave and crossed quadrilaterals?
Hint for the proof: Use Euclid's Proposition VI.2 (page 8).

*6.2.3. Construct a convex quadrilateral and its associated midpoint quadrilateral. Calculate the area of each. What is the relationship between the areas? Find analogous relationships for concave and crossed quadrilaterals.

The theorem is named for the French mathematician Pierre Varignon (1654–1722). The area relationship in the last exercise is often stated as part of Varignon's theorem.

6.3 Cyclic quadrilaterals

We have seen that every triangle can be circumscribed. In contrast, the vertices of a quadrilateral do not necessarily lie on a circle. To see this, observe that the first three vertices determine a unique circle and that the fourth vertex may or may not lie on it. Quadrilaterals whose vertices lie on a circle are quite special and will be useful in the next chapter.

Definition. A *cyclic quadrilateral* is a quadrilateral that is convex and whose vertices lie on a circle.

The requirement that a cyclic quadrilateral be convex guarantees that the order of the vertices on the quadrilateral matches the order of the vertices on the circle. The vertices of a concave quadrilateral cannot lie on a circle, so the requirement rules out crossed quadrilaterals.

Definition. A quadrilateral $\square ABCD$ is *inscribed* in the circle γ if all the vertices of $\square ABCD$ lie on γ.

Exercises

*6.3.1. Construct a circle γ and four points A, B, C, and D (in cyclic order) on γ. Measure angles $\angle ABC$ and $\angle CDA$ and calculate the sum of the measures. Do the same with angles $\angle BCD$ and $\angle DAB$. What relationship do you observe?

6.3.2. Prove the following theorem.

Euclid's Proposition III.22. *If $\square ABCD$ is a convex quadrilateral inscribed in the circle γ, then the opposite angles are supplements; i.e.,*

$$\mu(\angle ABC) + \mu(\angle CDA) = 180^\circ = \mu(\angle BCD) + \mu(\angle DAB).$$

Hint: Divide the angles using diagonals of the quadrilateral and apply the inscribed angle theorem.

6.3.3. Prove the following theorem.

Converse to Euclid's Proposition III.22. *If $\square ABCD$ is a convex quadrilateral such that the opposite angles are supplements, then $\square ABCD$ is a cyclic quadrilateral.*

Hint: Let γ be the circumscribed circle for $\triangle ABC$. Locate a point $D' \neq B$ such that D' lies on $\overrightarrow{BD}$ and γ. Show that the assumption $D \neq D'$ leads to a contradiction.

*6.3.4. Construct a circle γ and four points A, B, C, and D (in cyclic order) on γ. Let a, b, c, and d denote the lengths of the sides of $\square ABCD$. Define $s = (1/2)(a + b + c + d)$. Verify that

$$\alpha(\square ABCD) = \sqrt{(s-a)(s-b)(s-c)(s-d)}.$$

Is this formula for the area correct if $\square ABCD$ is not cyclic?

The formula in the last exercise is known as *Brahmagupta's formula*. It is named for the Indian mathematician Brahmagupta who discovered it in the seventh century. Heron's formula can be viewed as the special case in which $d = 0$.

6.4 Diagonals

Before you work the following exercises, you should review Exercise 1.3.8.

Exercises

6.4.1. Prove that a quadrilateral is a parallelogram if and only if the diagonals bisect each other.

6.4.2. Prove that a quadrilateral is a rhombus if and only if the diagonals are perpendicular and bisect each other.

6.4.3. Prove that a quadrilateral is a rectangle if and only if the diagonals are congruent and bisect each other.

6.4.4. Complete the following sentence: A quadrilateral is a square if and only if the diagonals ....

7

The Nine-Point Circle

One of the most remarkable discoveries in nineteenth century Euclidean geometry is that there is one circle that contains nine significant points associated with a triangle. In 1765 Euler proved that the midpoints of the sides and the feet of the altitudes of a triangle lie on a single circle. In other words, the medial and orthic triangles share the same circumcircle. Furthermore, the center of the common circumcircle lies on the Euler line of the original triangle. More than fifty years later, in 1820, Charles-Julien Brianchon (1783–1864) and Jean-Victor Poncelet (1788–1867) proved that the midpoints of the segments joining the orthocenter to the vertices lie on the same circle. As a result, the circle became known as the "nine-point circle." Later, Karl Wilhelm Feuerbach (1800–1834) proved that the nine-point circle has the additional property that it is tangent to all four of the equicircles; for this reason Feuerbach's name is often associated with the nine-point circle.

7.1 The nine-point circle

Let us begin with a statement of the theorem.

Nine-point Circle Theorem. *If $\triangle ABC$ is a triangle, then the midpoints of the sides of $\triangle ABC$, the feet of the altitudes of $\triangle ABC$, and the midpoints of the segments joining the orthocenter of $\triangle ABC$ to the three vertices of $\triangle ABC$ all lie on a single circle.*

The circle in Theorem 7.1 is the *nine-point circle* for $\triangle ABC$.

Exercises

*7.1.1. Construct a triangle and the nine points indicated in the theorem. Verify that they all lie on a circle, regardless of the shape of the triangle.

*7.1.2. Create a tool that constructs the nine-point circle for a triangle. The tool should accept the vertices of the triangle as input objects and should return both the circle and the nine points as output objects. Label all the points as in Figure 7.1. In particular, the vertices of the triangle are A, B, and C, the midpoints of the sides are D, E, and F as before, the feet of the altitudes are A', B', and C', the orthocenter is H, and the

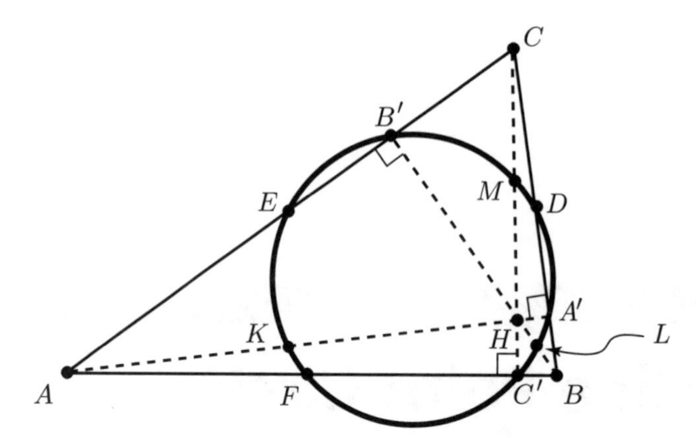

Figure 7.1. The nine-point circle

midpoints of the segments from H to the vertices of the triangle are K, L, and M. Experiment with triangles of different shapes to get a feel for the nine-point circle. Is it possible for the nine-point circle to be contained inside $\triangle ABC$? What is true of $\triangle ABC$ in that case?

Three points determine a circle, so the strategy for proving the nine-point circle theorem is to start with the circle determined by three of the points and prove that the other six points also lie on it. Specifically, let $\triangle ABC$ be a triangle and let D, E, and F be the midpoints of the sides of $\triangle ABC$. For the remainder of this section we will use γ to denote the circumcircle for the medial triangle $\triangle DEF$. The midpoints of the sides of the triangle obviously lie on γ. We will first show that the feet of the altitudes lie on γ and then we will show that the midpoints of the segments joining the orthocenter to the three vertices lie on γ.

Exercises

*7.1.3. Go back to your GeoGebra sketch from Exercise 7.1.2. Move the vertices A, B, and C so that your diagram looks as much like Figure 7.1 as possible. Use angle measurement and calculation to verify that each of the quadrilaterals $\square DEFA'$, $\square EFDB'$, and $\square DEFC'$ satisfies the following conditions:

(a) one pair of opposite sides is parallel,

(b) the other pair of opposite sides is congruent, and

(c) both pairs of opposite angles are supplements.

*7.1.4. When the vertices of $\triangle ABC$ are moved so that the diagram changes, there continue to be three trapezoids that satisfy the conditions of the previous exercise. But the order of the vertices must be adjusted. Try to understand the pattern before proceeding.

7.1.5. Assume $\triangle ABC$ is a triangle such that C' lies between F and B (see Figure 7.1). Prove that $\square DEFB$ is a parallelogram. Use the converse to Thales' theorem to prove that $DC' = DB$. Prove that $\square DEFC'$ is a cyclic quadrilateral. Conclude that C' always lies on the circle determined by D, E, and F.

*7.1.6. Go back to your GeoGebra sketch of the nine-point circle and move the vertices of the triangle $\triangle ABC$ to determine the other possible locations for C' on the line $\overleftrightarrow{AB}$. Modify the argument in the preceding exercise to conclude that in every case C' lies on the circle determined by D, E, and F.

The exercises prove that C' lies on the circumcircle of $\triangle DEF$. We could use the same argument (with the labels of points changed appropriately) to prove that A' and B' lie on the circle as well. We can therefore conclude that the feet of all three of the altitudes lie on the circle γ.

The next set of exercises will complete the proof of the nine-point circle theorem by showing that the remaining three points lie on γ as well. We continue to use the notation of Figure 7.1.

Exercises

7.1.7. Prove that $\overleftrightarrow{EM} \parallel \overleftrightarrow{AH}$ and conclude that $\overleftrightarrow{EM} \perp \overleftrightarrow{BC}$. Prove that $\overleftrightarrow{EF} \parallel \overleftrightarrow{BC}$ and conclude that $\overleftrightarrow{EM} \perp \overleftrightarrow{EF}$. Prove that E lies on the circle with diameter $\overline{MF}$.
Hint: Use the Converse to Thales' theorem (Exercise 0.10.4).

7.1.8. Prove that C' also lies on the circle with diameter $\overline{MF}$.

The two exercises show that the circle with diameter $\overline{MF}$ contains all four of the points E, F, C', and M. By uniqueness of the circumcircle, there is only one circle that contains the three points E, F, and C'. Since γ is a circle that contains E, F, and C', it is that unique circle and must be the circle that has $\overline{MF}$ as diameter. Thus we can conclude that M lies on γ. Similar arguments show that K and L lie on γ, so the proof of the nine-point circle theorem is complete.

We have actually proved more than is stated in the nine-point circle theorem—not only do the points K, L, and M lie on the nine-point circle, but $\overline{KD}$, $\overline{LE}$, and $\overline{MF}$ are diameters of the circle.

7.2 The nine-point center

It should be clear from Figure 7.1 that the orthocenter H is not the center of the nine-point circle. In fact the center of the nine-point circle is a new triangle center that we have not encountered before.

Definition. The center of the nine-point circle is called the *nine-point center* of $\triangle ABC$. It is denoted by N.

Exercises

*7.2.1. Create a tool that constructs the nine-point center of a triangle.
Hint: the nine-point center is the circumcenter of the medial triangle, so you should be able to make this tool by combining two others.

*7.2.2. Construct a triangle, its circumcenter, its orthocenter, and its nine-point center. Verify that the nine-point center is the midpoint of the segment joining the circumcenter to the orthocenter.

Here is a statement of the theorem you verified in the last exercise.

Nine-Point Center Theorem. *The nine-point center is the midpoint of the line segment from the circumcenter to the orthocenter.*

In particular, the nine-point center lies on the Euler line. We now know a total of five points on the Euler line: the centroid, the circumcenter, the orthocenter, the de Longchamps point, and the nine-point center.

Exercises

7.2.3. Prove the nine-point center theorem.
Hint: Use Figure 7.2 and the secant line theorem.

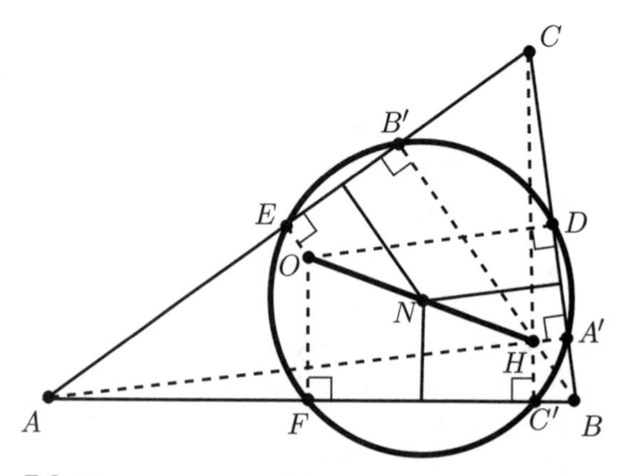

Figure 7.2. The orthocenter, the circumcenter, and the nine-point center

7.3 Feuerbach's theorem

Well after the nine-point circle had been discovered, Feuerbach proved the amazing result that the nine-point circle is tangent to each of the four equicircles.

Feuerbach's Theorem. *The nine-point circle is tangent to each of the four equicircles.*

Exercises

*7.3.1. Construct a triangle $\triangle ABC$ and use tools from your toolbox to construct the incircle, the three excircles, and the nine-point circle. Verify that the nine-point circle is tangent to each of the others. Explore different shapes for $\triangle ABC$ to determine possible configurations for the five circles.

*7.3.2. The *Feuerbach point* of the triangle is the point of tangency of the nine-point circle and the incircle. Construct the Feuerbach point for your triangle.

*7.3.3. The *Feuerbach triangle* is the triangle whose vertices are the three points of tangency of the nine-point circle with the three excircles. Construct the Feuerbach triangle for your triangle.

We will not attempt a proof of Feuerbach's theorem, but a proof may be found in §5.6 of [3].

8

Ceva's Theorem

In this chapter we finally complete the proofs of the concurrency theorems that were explored in earlier chapters. We will establish a theorem of Giovanni Ceva (1647–1734) that gives a necessary and sufficient condition for three lines through the vertices of a triangle to be concurrent and then we will derive all of the concurrency theorems as corollaries. We could have given *ad hoc* proofs of each of the concurrency results separately, but it seems better to expose the unifying principle behind them. One satisfying aspect of doing things this way is that the proof of Ceva's general theorem is not much more difficult than the proof of any one of the special concurrency results.

8.1 Exploring Ceva's theorem

All the concurrency theorems studied in previous chapters have common elements: they all address special cases of the following general problem.

Concurrency Problem. *Let $\triangle ABC$ be a triangle and let ℓ, m, and n be three lines such that A lies on ℓ, B lies on m, and C lies on n. Find necessary and sufficient conditions under which ℓ, m, and n are concurrent.*

The lines through the vertices can either be parallel to their opposite sidelines or intersect the opposite sidelines. We begin by looking at the two possibilities separately. Those lines that intersect the opposite sideline are more common and have a special name.

Definition. Let $\triangle ABC$ be a triangle and let L, M, and N be points on the sidelines $\overleftrightarrow{BC}$, $\overleftrightarrow{AC}$, and $\overleftrightarrow{AB}$, respectively. The lines $\ell = \overleftrightarrow{AL}$, $m = \overleftrightarrow{BM}$, and $n = \overleftrightarrow{CN}$ are called *Cevian lines* (or simply *cevians*) for $\triangle ABC$. A Cevian line is *proper* if it passes through exactly one vertex of $\triangle ABC$.

A Cevian line is specified by naming the vertex it passes through along with the point at which it intersects the opposite sideline. Thus, for example, the assertion that $\overleftrightarrow{AL}$ is a Cevian line for $\triangle ABC$ is understood to mean that L is a point on $\overleftrightarrow{BC}$.

Exercise 5.3.3 provides a hint about what happens when three proper Cevian lines are concurrent. As we saw there, the following quantity appears to be important:

$$d = \frac{AN}{NB} \cdot \frac{BL}{LC} \cdot \frac{CM}{MA}.$$

This quantity is defined and is positive provided all three Cevian lines are proper.

In the next set of exercises you will do some GeoGebra exploration that will further illuminate the significance of d and clarify the statement of the theorem we seek. To prepare for them, construct three noncollinear points A, B, and C, the three sidelines of $\triangle ABC$, and movable points L, M, and N on the sidelines $\overleftrightarrow{BC}$, $\overleftrightarrow{AC}$, and $\overleftrightarrow{AB}$, respectively. Use the Input Bar and Algebra View to calculate d.

It is a little tricky to get GeoGebra to display a nice formula for d in the Graphics View, but it can be done. To do so, open a text box, type

```
$\frac{AN}{NB}\cdot\frac{BL}{LC}\cdot\frac{CM}{MA} = $
```

and click on the value of d in the algebra view. Finally click on **LaTeX formula** and **OK**. The formula inside the \$-signs above is the LaTeX expression for the formula for d. When you type it as instructed, you should see something like Figure 8.1 in the Graphics View. You can then hide the Algebra View.

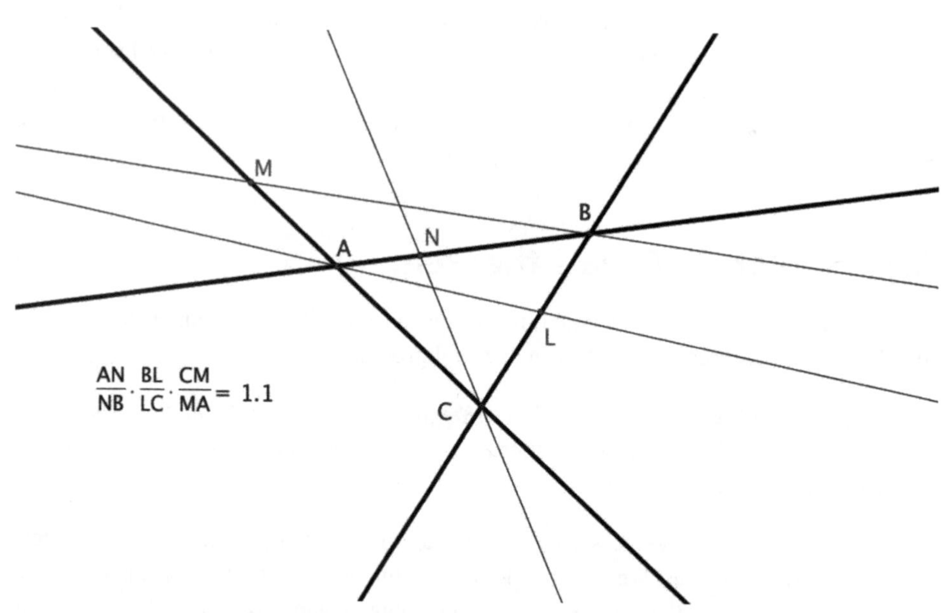

Figure 8.1. Formula for d in the GeoGebra Graphics View

It is assumed in all of the exercises below that $\triangle ABC$ and ℓ, m, and n are as in the statement of the Concurrency Problem.

Exercises

*8.1.1. Explore various positions for the vertices of the triangle and the points L, M, and N to verify the following result: *If three proper Cevian lines are concurrent, then* $d = 1$. Find examples of concurrent Cevian lines in which all three of the points L,

M, and N lie on the triangle $\triangle ABC$ and other examples in which at least one of the three points is not on $\triangle ABC$. Verify that $d = 1$ in all cases in which the proper Cevians are concurrent.

*8.1.2. Find an example in which $d = 1$ but the three Cevian lines are not concurrent.

*8.1.3. If the three Cevians are concurrent, how many of the points L, M, and N must lie on $\triangle ABC$? In your example for the preceding exercise, how many of the points were on the triangle?

*8.1.4. Find points L, M, and N such that the three Cevian lines are mutually parallel. (Use the Parallel Line tool to make sure the lines are exactly parallel, not just approximately parallel.) What is d equal to in this case? How many of the points L, M, and N lie on $\triangle ABC$?

*8.1.5. Let ℓ be the line through A that is parallel to the sideline $\overleftrightarrow{BC}$. Find points M and N so that ℓ is concurrent with the Cevian lines $\overleftrightarrow{BM}$ and $\overleftrightarrow{CN}$. When the three lines are concurrent, how many of the points M and N lie on the triangle? There is no point L in this case, but it is possible to compute the other two factors in d. Calculate

$$d' = \frac{AN}{NB} \cdot \frac{CM}{MA}$$

when the three lines are concurrent.

*8.1.6. Now let ℓ be the line through A that is parallel to the sideline $\overleftrightarrow{BC}$ and let m be the line through B that is parallel to the sideline $\overleftrightarrow{AC}$. Is there a Cevian line $n = \overleftrightarrow{CN}$ such that the three lines ℓ, m, and n are concurrent? If so, calculate AN/NB. Is N on the triangle?

*8.1.7. Is it possible for all three of the lines ℓ, m, and n to be parallel to the opposite sidelines and also concurrent?

*8.1.8. Suppose $\ell = \overleftrightarrow{AB}$. What must be true of m and n if the three lines are concurrent?

In the exercises you probably discovered that $d = 1$ if three proper Cevian lines are concurrent or if they are mutually parallel. But the converse does not hold: it is possible that $d = 1$ even if the lines are not concurrent or mutually parallel. It appears that the converse depends on how many of the points L, M, and N lie on $\triangle ABC$. You also saw that there are exceptional cases in which the lines ℓ, m, and n are not Cevian lines but are still concurrent. It would appear that the statement of a theorem that covers all the possibilities and all the special cases would have to be quite complicated. The next order of business is to extend the Euclidean plane and distance measurements in it in a way that allows us to give one succinct statement that summarizes all of these observations.

8.2 Sensed ratios and ideal points

This section contains several new ideas and definitions that will allow us to give a concise statement of the most general form of Ceva's theorem.

The first issue we address is how to include in d information about how many of the points L, M, and N are on the triangle and how many are not. We will accomplish that by adding a sign to each of the three ratios in the formula for d. Start with three distinct collinear points P, Q, and R. Since both distances involved are positive, the ratio PQ/QR is always positive. We will introduce a sensed ratio, which is sometimes positive and sometimes negative.

Definition. Assume P, Q, and R are three distinct collinear points. Define the *sensed ratio* **PQ/QR** by

$$\frac{\mathbf{PQ}}{\mathbf{QR}} = \begin{cases} \dfrac{PQ}{QR} & \text{if } Q \text{ is between } P \text{ and } R\text{, and} \\ -\dfrac{PQ}{QR} & \text{if } Q \text{ is not between } P \text{ and } R. \end{cases}$$

Boldface is used to denote the sensed ratio to distinguish it from the unsensed ratio.

It is possible to assign a direction to a line, so it makes sense to speak of *directed distances*. The directed distance **AB** between two points A and B on a line is considered to be positive if the direction from A to B agrees with the direction assigned to the line and is negative otherwise. Directed distances are not well defined because there are two ways to assign a direction to a given line. But reversing the direction of the line changes the sign of all the directed distances between points on the line. Thus the ratio of two directed distances is well defined because the same answer is obtained regardless of which direction is assigned to the line. The sensed ratio in the previous paragraph is the quotient of two directed distances.

When we state Ceva's theorem, we will replace the positive quantity d (defined in the last section) by a real number s, where s is the product of the sensed ratios. Specifically, we will define

$$s = \frac{\mathbf{AN}}{\mathbf{NB}} \cdot \frac{\mathbf{BL}}{\mathbf{LC}} \cdot \frac{\mathbf{CM}}{\mathbf{MA}}.$$

If all three of the points L, M, and N lie on the triangle $\triangle ABC$ or exactly one of the points lies on the triangle, then $s = d$. If either two of the points or none of the points L, M, and N lie on the triangle $\triangle ABC$, then s is negative and $s = -d$. Using s in the statement of the theorem will allow us to specify the numerical value of d and impose restrictions on the number of points that lie on the triangle at the same time.

Two other reasons Ceva's theorem is complicated to state in complete generality are that mutually parallel lines behave like concurrent lines and that a line through a vertex might be parallel to the opposite sideline and hence not a Cevian line. The correct setting for Ceva's theorem is the "extended Euclidean plane," in which any two lines intersect. We will define the extended Euclidean plane to consist of all points in the Euclidean plane together with some additional ideal points, called "points at infinity." The ideal points are added to the plane so that two parallel lines intersect at one of them.

Definition. The *extended Euclidean plane* consists of all the points in the ordinary Euclidean plane together with one additional point for each collection of mutually parallel lines in the plane. The points in the Euclidean plane will be called *ordinary points* and the new points are called *ideal points* or *points at infinity*. The set of all ideal points is called the *line at infinity*.

Intuitively what we have done is added one point to each line in the plane. This point is "at infinity" because it is infinitely far away from any of the ordinary points. We can think of it as being a point out at the end of the line. Since there is only one ideal point on each line we approach the same ideal point on a line by traveling towards either of the two ends of the line. We will not attempt to picture the ideal points in our diagrams since they are infinitely far away and thus beyond what we can see. They complete the ordinary plane so that it is not necessary to mention exceptional special cases in the statements of theorems. For example, every pair of distinct lines in the extended Euclidean plane intersect in exactly one point, with no exceptions. If the two lines are parallel in the ordinary plane, they share an ideal point and that is their point of intersection. If the two lines are not parallel in the ordinary plane, then they have different ideal points but intersect at an ordinary point.

Here is a summary of the important properties of the ideal points.

- Each ordinary line has been extended to contain exactly one ideal point.
- Parallel ordinary lines share a common ideal point.
- Ordinary lines that are distinct and nonparallel have distinct ideal points.

In the extended plane every line through a vertex of an ordinary triangle is a Cevian line. To see this, consider an ordinary triangle $\triangle ABC$ and a line ℓ that passes through vertex A. Either ℓ intersects the sideline $\overleftrightarrow{BC}$ at a point L or $\ell \parallel \overleftrightarrow{BC}$. In the first case ℓ is obviously a Cevian line, but in the second case ℓ is also a Cevian line with L being the ideal point on $\overleftrightarrow{BC}$.

Now that we have added additional points to the plane, we must extend our definition of sensed ratio to include them. To understand the next definition, think of two fixed points A and B and a movable point C on the line $\overleftrightarrow{AB}$. As C approaches either end of the line, C is not between A and B so the sensed ratio $\mathbf{AC}/\mathbf{CB}$ is negative. Furthermore, the distances AC and CB both approach infinity. But the difference between AC and CB is constant, so they approach infinity at the same rate and their ratio has a limit of 1. In other words, if I is the ideal point on $\overleftrightarrow{AB}$, then

$$\lim_{C \to I} \frac{\mathbf{AC}}{\mathbf{CB}} = -1.$$

This reasoning justifies the following definition.

Definition. Let A and B be distinct ordinary points and let I be the ideal point on $\overleftrightarrow{AB}$. Define

$$\frac{\mathbf{AI}}{\mathbf{IB}} = -1.$$

Finally, we extend ordinary arithmetic of fractions to cover the possibility that a Cevian line might not be proper. (This can happen, for example, in the altitude concurrence theorem.) We will adopt the convention that $a/b = c/d$ provided $ad = bc$. Thus it is possible for b to be zero in the fraction a/b, but if $b = 0$ and $a/b = c/d$, then either $a = 0$ or $d = 0$. In particular, if we write

$$\frac{AN}{NB} \cdot \frac{BL}{LC} \cdot \frac{CM}{MA} = 1,$$

this does not necessarily imply that $LC \neq 0$. Instead it means that if $LC = 0$, then at least one of AN, BL, or CM must also be zero.

Exercises

*8.2.1. Construct two lines ℓ and m. Mark their point of intersection. Now move a point on one line so that the line rotates while the other line stays fixed. Watch what happens to the point of intersection as the lines pass the position at which they are parallel. Explain how what you observe justifies the choice of just one ideal point on each line even though the line has two ends.

*8.2.2. Construct a line and points A, B, and X on the line. Calculate the ratio AX/XB. (GeoGebra will only calculate an unsensed ratio so you will have to supply the sign.) For which points X on $\overleftrightarrow{AB}$ is this ratio defined? Watch what happens to the ratio as X moves along the line. Use your observations to draw a graph of the function defined by

$$f(X) = \frac{\mathbf{AX}}{\mathbf{XB}}.$$

Hint: f is a function whose domain consists of the points on $\overleftrightarrow{AB}$ (excluding B) and whose range consists of real numbers, so your graph should have the line $\overleftrightarrow{AB}$ as its horizontal axis and the real line as the vertical axis.

8.2.3. Let A and B be distinct points. Prove that for each real number $r \in (-\infty, \infty)$ there is exactly one point on the extended line $\overleftrightarrow{AB}$ such that $\mathbf{AX}/\mathbf{XB} = r$. Which point on $\overleftrightarrow{AB}$ does not correspond to any real number r?

8.2.4. Draw an example of a triangle in the extended Euclidean plane that has one ideal vertex. Is there a triangle in the extended plane that has two ideal vertices? Could there be a triangle with three ideal vertices?

8.3 The standard form of Ceva's theorem

We are finally ready for the statement of Ceva's theorem.

Ceva's Theorem. *Let $\triangle ABC$ be an ordinary triangle. The Cevian lines $\overleftrightarrow{AL}$, $\overleftrightarrow{BM}$, and $\overleftrightarrow{CN}$ are concurrent if and only if*

$$\frac{\mathbf{AN}}{\mathbf{NB}} \cdot \frac{\mathbf{BL}}{\mathbf{LC}} \cdot \frac{\mathbf{CM}}{\mathbf{MA}} = +1.$$

The statement of the theorem assumes that the triangle is an ordinary triangle, so A, B, and C are ordinary points. There is no assumption that the points L, M, and N are ordinary, nor is there there an assumption that the point of concurrency is ordinary. In addition, L, M, and N may be at vertices of the triangle. Thus the theorem covers all possible lines through the vertices of the triangle and gives a complete solution to the Concurrency Problem stated at the beginning of the chapter.

Because we have stated Ceva's theorem in its ultimate generality, a complete proof will involve a number of special cases and this might make the proof appear to be complicated.

To make certain that the simplicity of the basic theorem is not lost, we will begin with a proof of the special case in which the point of concurrency is inside the triangle; this case covers most of the applications. It is the special case in which all three of the points L, M, and N are in the interior of the segments that form the sides of $\triangle ABC$. It follows from the crossbar theorem and related foundational results that the point of concurrency of the lines $\overleftrightarrow{AL}$, $\overleftrightarrow{BM}$, and $\overleftrightarrow{CN}$ is inside the triangle $\triangle ABC$ if and only if all three of the points L, M, and N are in the interiors of the segments that form the sides of $\triangle ABC$. That fact will be assumed in this section. The proof is outlined in the next two exercises.

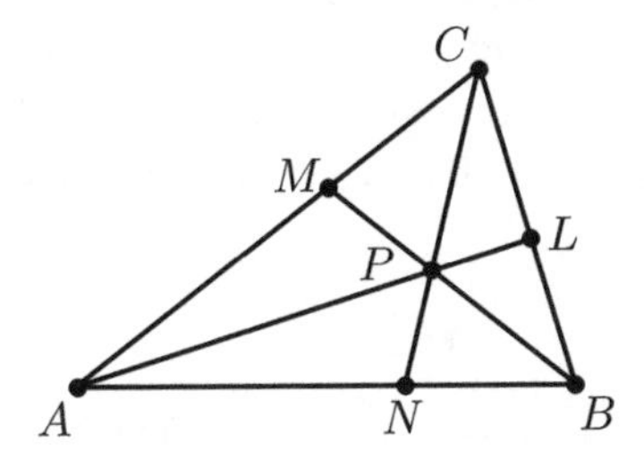

Figure 8.2. A special case of Ceva's theorem

Exercises

8.3.1. Let $\triangle ABC$ be a triangle and let L, M, and N be points in the interiors of the sides $\overline{BC}$, $\overline{AC}$, and $\overline{AB}$, respectively. Prove that if the three Cevian lines $\overleftrightarrow{AL}$, $\overleftrightarrow{BM}$, and $\overleftrightarrow{CN}$ are concurrent, then

$$\frac{\mathbf{AN}}{\mathbf{NB}} \cdot \frac{\mathbf{BL}}{\mathbf{LC}} \cdot \frac{\mathbf{CM}}{\mathbf{MA}} = 1.$$

Hint: Let k be the line through A that is parallel to $\overleftrightarrow{BC}$, let R be the point at which $\overleftrightarrow{BM}$ intersects k, and let S be the point at which $\overleftrightarrow{CN}$ intersects k. Use similar triangles to show that $AN/NB = SA/CB$, $CM/MA = CB/AR$, and $BL/LC = AR/SA$. The result follows by algebra.

8.3.2. Let $\triangle ABC$ be a triangle and let L, M, and N be points in the interiors of the sides $\overline{BC}$, $\overline{AC}$, and $\overline{AB}$, respectively. Prove that if

$$\frac{\mathbf{AN}}{\mathbf{NB}} \cdot \frac{\mathbf{BL}}{\mathbf{LC}} \cdot \frac{\mathbf{CM}}{\mathbf{MA}} = 1,$$

then the three Cevian lines $\overleftrightarrow{AL}$, $\overleftrightarrow{BM}$, and $\overleftrightarrow{CN}$ are concurrent.
Hint: By the crossbar theorem there is a point P at which $\overline{AL}$ intersects $\overline{BM}$. Choose N' to be the point at which $\overrightarrow{CP}$ intersects $\overline{AB}$. Note that $\overleftrightarrow{AL}$, $\overleftrightarrow{BM}$, and $\overleftrightarrow{CN'}$ are concurrent. Use the part of the theorem you have already proved and Exercise 8.2.3 to prove that $N = N'$.

We now proceed to prove Ceva's theorem in general. We will begin by proving that if the three lines are concurrent, then $s = 1$. Once that has been established, the proof used in the last exercise can easily be modified to prove the converse.

Assume in the following exercises that $\triangle ABC$ is a triangle, the lines $\overleftrightarrow{AL}$, $\overleftrightarrow{BM}$, and $\overleftrightarrow{CN}$ are Cevian lines for $\triangle ABC$, and

$$s = \frac{\mathbf{AN}}{\mathbf{NB}} \cdot \frac{\mathbf{BL}}{\mathbf{LC}} \cdot \frac{\mathbf{CM}}{\mathbf{MA}}.$$

We assume that A, B, and C are ordinary points in the Euclidean plane, but L, M, and N are points in the extended plane. Thus one or more of the points L, M, and N may be ideal, in which case the corresponding factors in s are -1. It is also possible that one or more of L, M, and N may be a vertex of $\triangle ABC$. In that case there might be a 0 in the denominator of s. When that happens the factors of s should not be considered separately, but the entire numerator and the entire denominator should be considered. If one of the factors in the denominator of s is 0, then $s = -1$ means that there is also a factor in the numerator that is equal to 0.

Exercises

8.3.3. Prove that if $\overleftrightarrow{AL}$, $\overleftrightarrow{BM}$, and $\overleftrightarrow{CN}$ are proper Cevian lines that are concurrent at an ordinary point P, then either all three of the points L, M, and N lie on $\triangle ABC$ or exactly one of them does.
Hint: The three sidelines divide the exterior of $\triangle ABC$ into six regions. Consider the possibility that P lies in each of them separately. Don't forget that one or more of the points L, M, and N might be ideal.

8.3.4. Prove that if $\overleftrightarrow{AL}$, $\overleftrightarrow{BM}$, and $\overleftrightarrow{CN}$ are proper Cevian lines that are concurrent at an ideal point, then exactly one of the three points L, M, and N lies on $\triangle ABC$.

8.3.5. Prove that if $\overleftrightarrow{AL}$, $\overleftrightarrow{BM}$, and $\overleftrightarrow{CN}$ are proper Cevian lines that are concurrent at an ordinary point P and all three of L, M, and N are ordinary, then $s = 1$.
Hint: The case in which all three of the points L, M, and N lie on $\triangle ABC$ was covered in Exercise 8.3.1, so you may assume that only M lies on $\triangle ABC$. Proceed as in the proof of Exercise 8.3.1; there are two possible diagrams, both different from that in Exercise 8.3.1, but you should still be able to find the similar triangles you need.

8.3.6. Prove that if $\overleftrightarrow{AL}$, $\overleftrightarrow{BM}$, and $\overleftrightarrow{CN}$ are proper Cevian lines that are concurrent at an ideal point and all three of L, M, and N are ordinary, then $s = 1$.
Hint: Concurrent at an ideal point means parallel. Essentially the same proof works again.

8.3.7. Prove that if $\overleftrightarrow{AL}$, $\overleftrightarrow{BM}$, and $\overleftrightarrow{CN}$ are proper Cevian lines that are concurrent at an ordinary point P and L is ideal but M and N are ordinary, then $s = 1$.
Hint: Take another look at your sketch and your conclusions in Exercise 8.1.5.

8.3.8. Prove that if $\overleftrightarrow{AL}$, $\overleftrightarrow{BM}$, and $\overleftrightarrow{CN}$ are proper Cevian lines that are concurrent at an ordinary point P and L and M are ideal but N is ordinary, then $s = 1$.
Hint: Take another look at your sketch and your conclusions in Exercise 8.1.6.

8.3.9. Prove that at least one of L, M, and N must be ordinary if the proper Cevian lines $\overleftrightarrow{AL}$, $\overleftrightarrow{BM}$, and $\overleftrightarrow{CN}$ are concurrent.

8.3.10. Prove that if $L = B$ (so that $\overleftrightarrow{AL}$ is the sideline $\overleftrightarrow{AB}$ of the triangle) and the three Cevians $\overleftrightarrow{AL}$, $\overleftrightarrow{BM}$, and $\overleftrightarrow{CN}$ are concurrent, then either $\overleftrightarrow{BM} = \overleftrightarrow{AB}$ or $\overleftrightarrow{CN} = \overleftrightarrow{BC}$. Prove that $s = 1$ in either case.

8.3.11. Check that Exercises 8.3.3 through 8.3.10 prove $s = 1$ in every case in which the Cevian lines $\overleftrightarrow{AL}$, $\overleftrightarrow{BM}$, and $\overleftrightarrow{CN}$ are concurrent.

8.3.12. Prove that if $s = 1$, then the Cevian lines $\overleftrightarrow{AL}$, $\overleftrightarrow{BM}$, and $\overleftrightarrow{CN}$ are concurrent. Hint: Proceed as in Exercise 8.3.2. Be sure to check that this proof works in every case.

8.4 The trigonometric form of Ceva's theorem

There is second form of Ceva's theorem that is sometimes more convenient to apply than the standard form. It expresses the concurrence criterion in terms of sines of angles rather than distances, so it is useful for applications in which it is simpler to measure the angles than the lengths.

As in the standard form, sense must be taken into account in the concurrence condition. For the trigonometric form this is a matter of using directed measures for the angles. The measure of an angle $\angle BAC$ is positive if it is measured in the counterclockwise direction from $\overrightarrow{AB}$ to $\overrightarrow{AC}$ and it is negative if the rotation from $\overrightarrow{AB}$ to $\overrightarrow{AC}$ is clockwise. (This is the same way angles are measured in calculus.) We will use boldface for the angle when it is to be considered a directed angle. Thus $\angle\mathbf{BAC}$ denotes the directed angle from $\overrightarrow{AB}$ to $\overrightarrow{AC}$.

Trigonometric Form of Ceva's Theorem. *Let $\triangle ABC$ be an ordinary triangle. The Cevian lines $\overleftrightarrow{AL}$, $\overleftrightarrow{BM}$, and $\overleftrightarrow{CN}$ are concurrent if and only if*

$$\frac{\sin(\angle\mathbf{BAL})}{\sin(\angle\mathbf{LAC})} \cdot \frac{\sin(\angle\mathbf{CBM})}{\sin(\angle\mathbf{MBA})} \cdot \frac{\sin(\angle\mathbf{ACN})}{\sin(\angle\mathbf{NCB})} = +1.$$

The proof is based on the following simple lemma.

Lemma. *If $\triangle ABC$ is an ordinary triangle and L is a point on $\overleftrightarrow{BC}$, then*

$$\frac{\mathbf{BL}}{\mathbf{LC}} = \frac{AB \cdot \sin(\angle\mathbf{BAL})}{AC \cdot \sin(\angle\mathbf{LAC})}.$$

Proof. We will prove the theorem in case L is between B and C (see Figure 8.3) and leave the general case as an exercise. Let h denote the height of triangle $\triangle ABC$. Then

$$\frac{\mathbf{BL}}{\mathbf{LC}} = \frac{(1/2)\,h \cdot BL}{(1/2)\,h \cdot LC} = \frac{(1/2)\,AB \cdot AL \cdot \sin(\angle BAL)}{(1/2)\,AL \cdot AC \cdot \sin(\angle LAC)} = \frac{AB \cdot \sin(\angle\mathbf{BAL})}{AC \cdot \sin(\angle\mathbf{LAC})}.$$

The first equation is just algebra, the second equation is based on two applications of Exercise 0.11.1, and the third equation uses the fact that L is between B and C so the two angles have the same direction. □

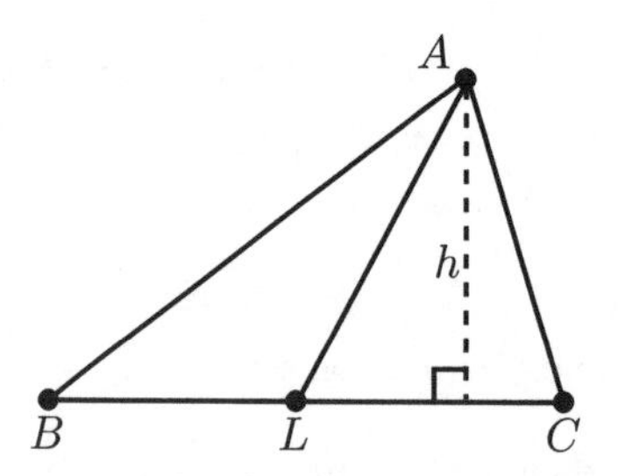

Figure 8.3. Proof of the lemma

Exercises

8.4.1. Modify the proof above to cover the remaining cases of the lemma.
Hint: L might equal B or C, or L might be outside $\overline{BC}$.

8.4.2. Use the lemma to prove the trigonometric form of Ceva's theorem.

8.5 The concurrence theorems

In this section we apply Ceva's theorem to prove the concurrence results that we have encountered thus far in the course. Before getting to those proofs, however, we make one observation about applying Ceva's theorem.

When we want to apply Ceva's theorem we will usually not find the triangle and the Cevian lines conveniently labeled with exactly the letters that were used in the statement of Ceva's theorem. The pattern in the expression

$$\frac{\mathbf{AN}}{\mathbf{NB}} \cdot \frac{\mathbf{BL}}{\mathbf{LC}} \cdot \frac{\mathbf{CM}}{\mathbf{MA}}$$

is that the vertices of the triangle are listed in cyclic order around the triangle. The first fraction involves the two endpoints of the first side of the triangle and the point on the corresponding sideline. The second fraction involves the second side of the triangle, oriented so that the second side begins at the point where the first segment ended, and so on. When you apply the theorem you may begin at any vertex of the triangle and you may go around the triangle in either direction. But once you select a starting vertex and a direction around the triangle, you must follow the rest of the pattern.

Exercises

8.5.1. **Existence of the centroid.** Prove that the medians of a triangle are concurrent.

8.5.2. **Existence of the orthocenter.** Prove that the altitudes of a triangle are concurrent.
Hint: Use the trigonometric form of Ceva's theorem. For each angle in the numerator, find a congruent angle in the denominator. Be careful with the special case of a right triangle.

8.5.3. **Existence of the circumcenter.** Prove that the perpendicular bisectors of the sides of a triangle are concurrent.
Hint: Be sure to notice that the circumcenter breaks the pattern: it is not the point of concurrence of Cevian lines. Use Exercise 5.1.4.

8.5.4. Existence of the incenter. Prove that the interior angle bisectors of a triangle are concurrent.

8.5.5. Existence of excenters. Prove that the bisector of an interior angle of a triangle and the bisectors of the remote exterior angles are concurrent.

8.5.6. Existence of the Gergonne point. Let L, M, and N be the points at which the incircle touches the sides of $\triangle ABC$. Prove that $\overline{AL}$, $\overline{BM}$, and $\overline{CN}$ are concurrent.
Hint: Use the external tangents theorem.

8.5.7. Existence of the Nagel point. Let $\triangle ABC$ be a triangle, let T_A be the point at which the A-excircle is tangent to $\overline{BC}$, let T_B be the point at which the B-excircle is tangent to $\overline{AC}$, and let T_C be the point at which the C-excircle is tangent to $\overline{AB}$. Prove that the segments $\overline{AT_A}$, $\overline{BT_B}$, and $\overline{CT_C}$ are concurrent.
Hint: Use Exercise 4.5.2 to prove that the ratios in this problem are the reciprocals of the ratios in Exercise 8.5.6.

8.5.8. Existence of the Vecten point. Prove that the three lines joining the vertices of a triangle to the centers of the outer squares on the opposite sides are concurrent.
Hint: Use Exercise 3.3.10.

8.6 Isotomic and isogonal conjugates and the symmedian point

Ceva's theorem allows us to define two interesting transformations of the plane associated with a triangle. Both are called *conjugates* because applying them twice results in the identity transformation. This is the same sense in which the word "conjugate" is used in connection with complex numbers. In other contexts a transformation with this property would be called an *involution*.

Let $\triangle ABC$ be an ordinary triangle and let P be a point in the extended plane. For simplicity we will assume that P does not lie on any of the sidelines of the triangle. In that case P is the point of concurrence of three proper Cevian lines $\overleftrightarrow{AL}$, $\overleftrightarrow{BM}$, and $\overleftrightarrow{CN}$.

Definition. The *isotomic conjugate* of P is the point of concurrence of the lines $\overleftrightarrow{AL'}$, $\overleftrightarrow{BM'}$, and $\overleftrightarrow{CN'}$, where L' is the reflection of L across the perpendicular bisector of $\overline{BC}$,

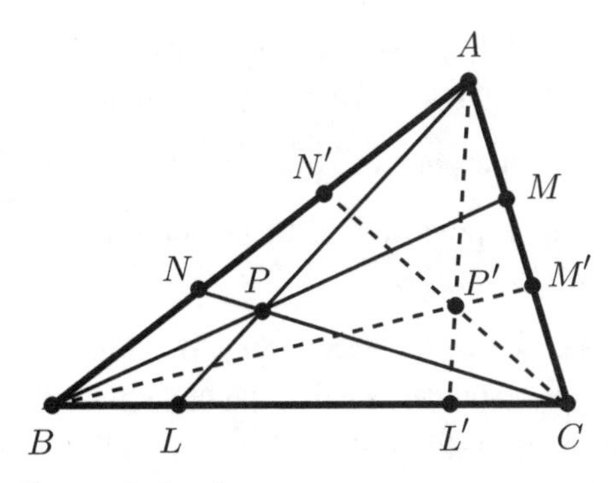

Figure 8.4. P' is the isotomic conjugate of P

M' is the reflection of M across the perpendicular bisector of $\overline{AC}$, and N' is the reflection of N across the perpendicular bisector of $\overline{AB}$.

Definition. The *isogonal* of a Cevian line $\overleftrightarrow{AL}$ is the reflection of $\overleftrightarrow{AL}$ through the angle bisector of $\angle CAB$. Similarly the isogonal of a Cevian line $\overleftrightarrow{BM}$ is the reflection of $\overleftrightarrow{BM}$ through the bisector of $\angle ABC$ and the isogonal of a Cevian line $\overleftrightarrow{CN}$ is the reflection of $\overleftrightarrow{CN}$ through the bisector of $\angle BCA$. The *isogonal conjugate* of P is the point of concurrence of the isogonals of $\overleftrightarrow{AL}$, $\overleftrightarrow{BM}$, and $\overleftrightarrow{CN}$.

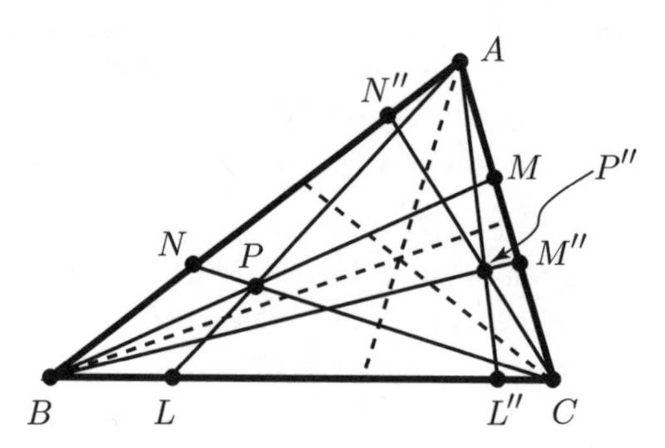

Figure 8.5. P'' is the isogonal conjugate of P

Both definitions presume that the new Cevian lines intersect; that is where Ceva's theorem comes in. The next two exercises show that the isotomic conjugate and the isogonal conjugate exist.

Exercises

8.6.1. Use the standard form of Ceva's theorem to prove that the three Cevian lines in the definition of isotomic conjugate are concurrent.

8.6.2. Use the trigonometric form of Ceva's theorem to prove that the three Cevian lines in the definition of isogonal conjugate are concurrent.

*8.6.3. Let $\triangle ABC$ be a triangle. Make a tool that constructs the isotomic conjugate of a point P that is not on any of the sidelines of the triangle.

*8.6.4. Let $\triangle ABC$ be a triangle. Make a tool that constructs the isogonal conjugate of a point P that is not on any of the sidelines of the triangle.

*8.6.5. Verify that the Gergonne point and the Nagel point are isotomic conjugates.

8.6.6. Prove that the Gergonne point and the Nagel point are isotomic conjugates.
Hint: Use Exercise 4.5.2.

*8.6.7. Verify that the incenter is, in general, the only point in the interior of a triangle that is its own isogonal conjugate. Find all points in the exterior of the triangle that are isogonal conjugates of themselves.

8.6.8. Prove that the incenter is its own isogonal conjugate.

8.6.9. Prove that the orthocenter and the circumcenter are isogonal conjugates.
Hint: Let γ be the circumcircle of $\triangle ABC$. Extend $\overline{AO}$ and $\overline{AH}$ until they intersect γ at points R and S as indicated in Figure 8.6. Let $\overline{PQ}$ be a diameter of γ that is parallel to $\overline{BC}$ and let T be a point on γ such that $\overline{ST}$ is a diameter (see Figure 8.6).

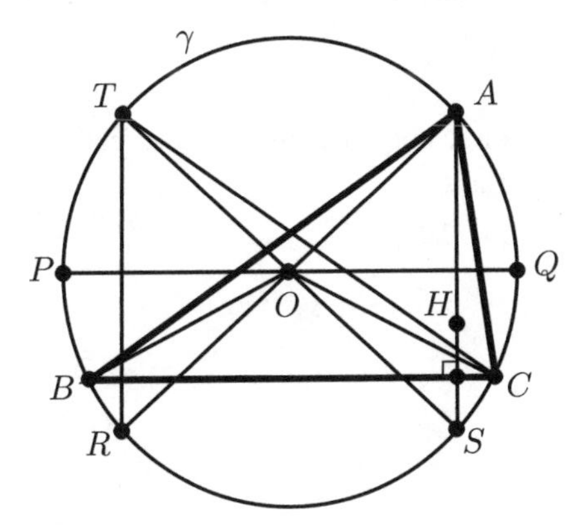

Figure 8.6. Proof that the circumcenter and the orthocenter are isogonal conjugates

Prove that $\overleftrightarrow{RT} \perp \overleftrightarrow{PQ}$. Prove that $\angle ROP \cong \angle POT \cong \angle QOS$ and prove that $\angle BOP \cong \angle COQ$. Conclude that $\angle BOR \cong \angle COS$. Apply Exercise 0.10.5 twice to conclude that $\angle BAR \cong \angle CTS$. Finally, use the inscribed angle theorem to conclude that $\angle BAO \cong \angle HAC$.

*8.6.10. Consider a triangle $\triangle ABC$ and a point P that does not lie on any of the sidelines of the triangle. Construct the three Cevians through P. What happens to the isogonals of the Cevians as P approaches the circumcircle of $\triangle ABC$? What is the isogonal conjugate of a point on the circumcircle?

8.6.11. Prove that the isogonal conjugate of a point on the circumcircle is an ideal point.
Hint: The isogonal conjugate of P is ideal if and only if the isogonals of the Cevians through P are parallel. Since the isogonals of the Cevians have already been proved concurrent, it is enough to show that two of them are parallel. Figure 8.7 shows one possible diagram. Apply Euclid's Proposition III.22 and the Angle sum theorem.

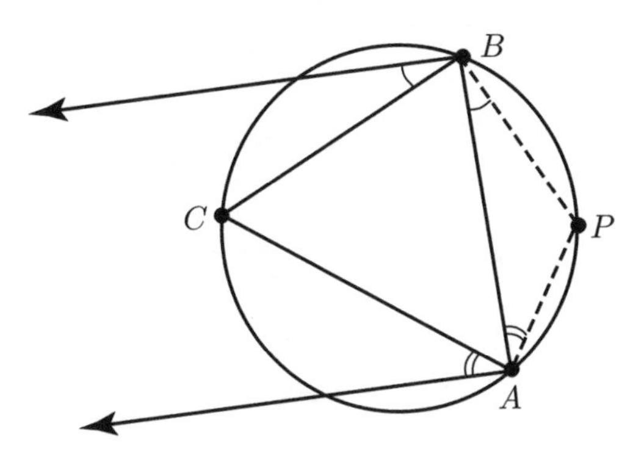

Figure 8.7. Proof that the isogonal conjugate of a point on the circumcircle is ideal

Definition. A *symmedian* for $\triangle ABC$ is the isogonal of a median of the triangle. The *symmedian point* of $\triangle ABC$ is the point of concurrence of the three symmedians; i.e., the symmedian is the isogonal conjugate of the centroid.

The symmedian point is another triangle center. It is usually denoted by K.

Exercise

*8.6.12. Verify that the isotomic conjugate of the orthocenter is the symmedian point of the anticomplementary triangle.

9

The Theorem of Menelaus

The theorem we study next is ancient, dating from about the year AD 100. It was originally discovered by Menelaus of Alexandria (70–130), but it did not become well known until it was rediscovered by Ceva in the seventeenth century. The theorem is powerful and has many interesting applications, some of which will be explored in later chapters.

9.1 Duality

In the last chapter we studied the problem of determining when three lines through the vertices of a triangle are concurrent. In this chapter we study the problem of determining when three points on the sidelines of a triangle are collinear. The relationship between the two problems is an example of *duality*. Before studying the main result of the chapter we will pause to consider the principle of duality because discerning larger patterns such as duality can lead to a deeper understanding of the theorems of geometry than studying each geometric result in isolation.

Roughly speaking, the principle of duality asserts that any true statement in geometry should remain true when the words point and line are interchanged. Just as two points lie on exactly one line, so two lines intersect in exactly one point; just as three points may be collinear, so three lines may be concurrent. The use of the term *incident* makes these statements even more symmetric. For example, two distinct points are incident with exactly one line and two distinct lines are (usually) incident with exactly one point.

Passing to the extended Euclidean plane allows us to eliminate the exceptions, so the extended plane is the natural setting for duality. Specifically, the statement that two distinct lines are incident with exactly one point is true without exception in the extended plane. This makes it completely dual to the statement that two distinct points are incident with exactly one line (Euclid's first postulate). Girard Desargues (1591–1661) was the first to make systematic use of ideal points and his work led eventually to the development of a whole new branch of geometry known as *projective geometry*. A later chapter will include a study of the theorem of Desargues, which is a beautiful result that relates two dual properties of triangles. Desargues's theorem is foundational in projective geometry, but we will not pursue that branch of geometry any further.

The triangle itself is self dual. It can be thought of as determined by three noncollinear points (the vertices) or by three nonconcurrent lines (the sidelines). When the latter view of the triangle is being emphasized, the triangle is referred to by its dual name *trilateral*. Vertices and sidelines (not sides) are dual aspects of the triangle, so they should be interchanged when theorems regarding triangles are being dualized. This is the reason it is natural to use the sidelines of a triangle, rather than its sides, in so many of the theorems in this book.

The theorems of Ceva and Menelaus will be our primary example of dual theorems. The theorem of Ceva gives a criterion that tells us when three lines through the vertices of a triangle are concurrent; the theorem of Menelaus tells us that essentially the same criterion can be used to determine when three points on the sidelines of a triangle are collinear. Apparently what happened historically is that Ceva rediscovered the theorem of Menelaus and then discovered his own theorem by applying the principle of duality.

The principle of duality was first formalized by Charles Julien Brianchon (1785–1864) when he applied it to a theorem of Blaise Pascal (1623–1663). The theorems of Pascal and Brianchon serve as a beautiful illustration of dual theorems. The theorem of Pascal asserts that if a hexagon is inscribed in a circle, then the three points determined by pairs of opposite sidelines are collinear. Brianchon's theorem asserts that if a hexagon is circumscribed about a circle, then the lines determined by pairs of opposite vertices are concurrent. We will study those theorems in a later chapter.

9.2 The theorem of Menelaus

To simplify the statements in this section, we make a definition.

Definition. Let $\triangle ABC$ be a triangle. Three points L, M, and N such that L lies on $\overleftrightarrow{BC}$, M lies on $\overleftrightarrow{AC}$, and N lies on $\overleftrightarrow{AB}$ are called *Menelaus points* for the triangle. We say that a Menelaus point is *proper* if it is not at any of the vertices of the triangle.

It will be assumed that the vertices of the triangle are ordinary points, but one or more of the Menelaus points may be ideal. Let us begin with some GeoGebra exploration of when three Menelaus points are collinear.

Exercises

*9.2.1. Construct a triangle $\triangle ABC$ and proper Menelaus points L, M, and N for $\triangle ABC$. Calculate the quantity

$$d = \frac{AN}{NB} \cdot \frac{BL}{LC} \cdot \frac{CM}{MA}.$$

Move the vertices of the triangle and the Menelaus points to verify that if the Menelaus points are collinear, then $d = 1$.

*9.2.2. Verify that

$$\frac{AN}{NB} \cdot \frac{CM}{MA} = 1$$

whenever L is an ideal point and $\overleftrightarrow{MN}$ is parallel to $\overleftrightarrow{BC}$.

*9.2.3. Find an example of a triangle and Menelaus points such that $d = 1$ even though the Menelaus points are not collinear.

*9.2.4. Construct a triangle $\triangle ABC$ and a line ℓ that does not pass through any of the vertices of the triangle. Determine the answer to the question: What are the possible numbers of points of intersection of ℓ with the triangle?

The exercises indicate that if the Menelaus points are collinear, then $d = 1$. But they also show that the converse does not hold. Again the number of Menelaus points that lie on the triangle itself is important, so we must use

$$s = \frac{\mathbf{AN}}{\mathbf{NB}} \cdot \frac{\mathbf{BL}}{\mathbf{LC}} \cdot \frac{\mathbf{CM}}{\mathbf{MA}},$$

the product of the sensed ratios, rather than d. All of this is probably expected since it is exactly analogous to what happened in the case of Ceva's theorem.

One difference between this situation and that of Ceva's theorem is that the number of Menelaus points that are expected to lie on the triangle is even, so the number of factors in s that are positive is even, leaving an odd number of negative factors. This is because a line in the plane that does not contain any of the vertices of the triangle $\triangle ABC$ will either miss the triangle entirely or will intersect it in exactly two points. You should have convinced yourself of that when you did your GeoGebra exploration in Exercise 9.2.4. Pasch's Axiom, which was assumed in Chapter 0, makes this precise. We can conclude that if the three Menelaus points are proper and concurrent, then either zero or two of them lie on the triangle. In that case either three or one of the sensed ratios in s will be negative, so s itself is negative.

We can now state Menelaus's theorem in full generality. Again we assume that the vertices of the triangle are ordinary points in the Euclidean plane, but we allow the possibility that one or more of the Menelaus points is ideal.

Theorem of Menelaus. *Let $\triangle ABC$ be an ordinary triangle. The Menelaus points L, M, and N for $\triangle ABC$ are collinear if and only if*

$$\frac{\mathbf{AN}}{\mathbf{NB}} \cdot \frac{\mathbf{BL}}{\mathbf{LC}} \cdot \frac{\mathbf{CM}}{\mathbf{MA}} = -1.$$

The next exercises outline a proof of Menelaus's theorem. It would be possible to prove the theorem as a corollary of Ceva's theorem, but that proof is no simpler than a proof based on similar triangles. Since it is no more difficult, we give a proof based only on elementary Euclidean geometry and do not use Ceva's theorem.

Exercises

9.2.5. Assume that all three Menelaus points are proper and ordinary. Prove that if the three Menelaus points are collinear, then $s = -1$.
Hint: Assume, first, that L and M lie on the triangle and N does not (see Figure 9.1). Drop perpendiculars from A, B, and C to the line $\overleftrightarrow{LM}$ and call the feet R, S, and T, respectively. Let $r = AR$, $s = BS$, and $t = CT$. Use similar triangles to express each of the sensed ratios **AN/NB**, **BL/LC**, and **CM/MA** in terms of r, s, and t and then use algebra to derive the Menelaus formula. Use GeoGebra to determine what the other possible figures look like and modify the argument to fit them as well.

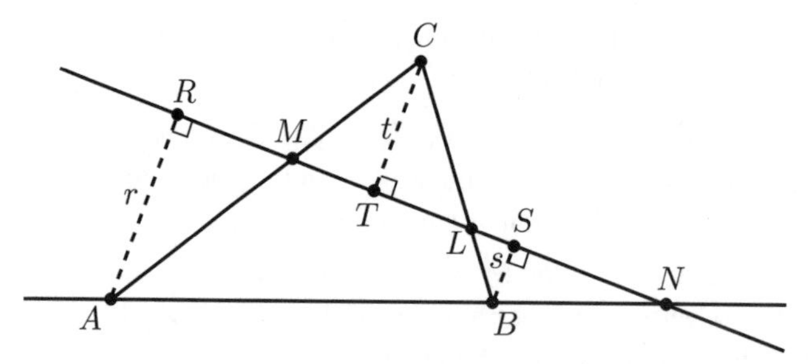

Figure 9.1. Proof of Menelaus's theorem in case two points lie on the triangle

9.2.6. Prove the theorem in case one or more of the Menelaus points is improper. Specifically, prove each of the following statements and then explain why this suffices to prove the theorem in every case in which at least one of the Menelaus points is improper.

(a) If $L = C$ and the three Menelaus points are collinear, then either $N = A$ or $M = C$ and (in either case) $s = -1$.

(b) If $L = C$ and $s = -1$, then either $N = A$ or $M = C$.

9.2.7. Assume that all three Menelaus points are proper and that exactly one of them is ideal. Prove that $s = -1$ if and only if the three Menelaus points are collinear.

9.2.8. Prove that it is impossible for the Menelaus points to be collinear if exactly two of them are ideal points. Prove that $s = -1$ in case all three Menelaus points are ideal. (In that case they are collinear because they all lie on the line at infinity.)

9.2.9. Prove the converse of Menelaus's theorem; i.e., prove that if $s = -1$, then the three Menelaus points are collinear.

There is also a trigonometric form of Menelaus's theorem. The proof is much like the proof of the trigonometric form of Ceva's theorem.

Trigonometric Form of the Theorem of Menelaus. *Let $\triangle ABC$ be an ordinary triangle. The Menelaus points L, M, and N for $\triangle ABC$ are collinear if and only if*

$$\frac{\sin(\angle \mathbf{BAL})}{\sin(\angle \mathbf{LAC})} \cdot \frac{\sin(\angle \mathbf{CBM})}{\sin(\angle \mathbf{MBA})} \cdot \frac{\sin(\angle \mathbf{ACN})}{\sin(\angle \mathbf{NCB})} = -1.$$

Exercise

9.2.10. Prove the trigonometric form of the theorem of Menelaus.

10

Circles and Lines

Before proceeding to the applications of Menelaus's theorem we develop some results about circles and lines that we will need.

10.1 The power of a point

We begin by defining the power of a point with respect to a circle.

Definition. Let β be a circle and let O be a point. Choose a line ℓ such that O lies on ℓ and ℓ intersects β. Define the *power of O with respect to β* to be

$$p(O, \beta) = \begin{cases} (\mathbf{OP})^2 & \text{if } \ell \text{ is tangent to } \beta \text{ at } P, \text{ or} \\ (\mathbf{OQ})(\mathbf{OR}) & \text{if } \ell \text{ intersects } \beta \text{ at two points } Q \text{ and } R. \end{cases}$$

The distances in the definition are directed distances, so the power of a point inside β is negative while the power of a point outside β is positive (and the power of a point on β is 0).

We will prove two of Euclid's most famous propositions. The first asserts that the power of a point is well defined because it does not depend on which line ℓ is used in the definition (provided O lies on the line and the line intersects β). The second gives a relationship between the angles of a triangle and the angles made by a line that is tangent to the circumcircle at a vertex. Both results will be useful in the next chapter.

Euclid's Proposition III.36. *The power of O with respect to β is well defined; i.e., the same numerical value for the power is obtained regardless of which line ℓ is used in the definition (provided O lies on ℓ and ℓ intersects β).*

Euclid's Proposition III.32. *Let $\triangle ABC$ be a triangle inscribed in the circle γ and let t be the line that is tangent to γ at A. If D and E are points on t such that A is between D and E and D is on the same side of $\overleftrightarrow{AC}$ as B, then*

$$\angle DAB \cong \angle ACB \quad \text{and} \quad \angle EAC \cong \angle ABC.$$

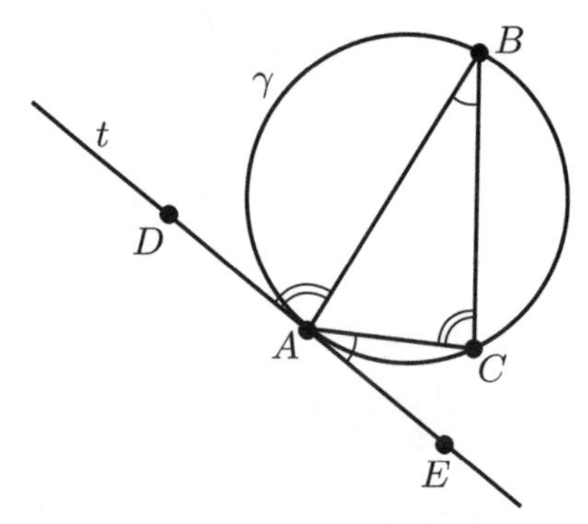

Figure 10.1. Proposition III.32: $\angle DAB \cong \angle ACB$ and $\angle EAC \cong \angle ABC$

Exercises

*10.1.1. Construct two points O and S and a circle β. Keep O fixed and move S so that $\overleftrightarrow{OS}$ that intersects β in two points. Mark the two points of intersection and label them Q and R. Measure OQ and OR and calculate $(OQ)(OR)$. Now move S (keeping O and β fixed) and observe that $(OQ)(OR)$ remains constant. Verify that this is true whether O is inside β or outside β.

10.1.2. Prove Euclid's Proposition III.36.
Hint: Let B be the center of β. First check the case in which $O = B$. Then assume $O \neq B$. Let S and T be the two points at which $\overleftrightarrow{OB}$ intersects β; see Figure 10.2. If ℓ intersects β in two points Q and R, use similar triangles to prove that $(OQ)(OR) = (OS)(OT)$. If ℓ is tangent to β at P, use the Pythagorean theorem to prove that $(OP)^2 = (OS)(OT)$. Explain why this is enough to prove the theorem.

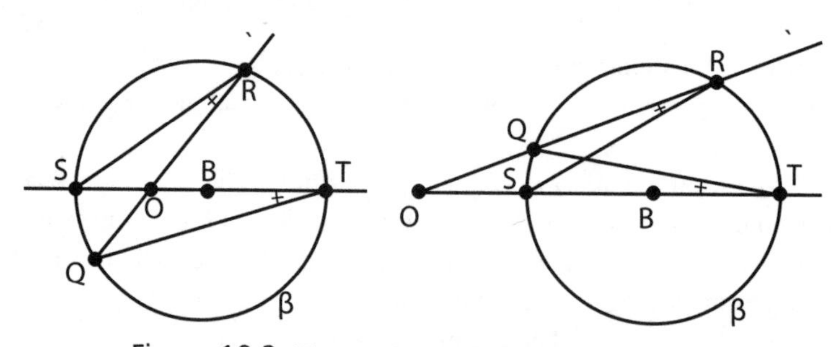

Figure 10.2. The proof of Euclid's Proposition III.36

*10.1.3. Construct a circle γ and three points A, B, and C on γ. Then construct the tangent line t at A. Measure the angles made by t with $\overrightarrow{AB}$ and $\overrightarrow{AC}$ and the interior angles of $\triangle ABC$. Find a relationship between the measures of the angles of $\triangle ABC$ and those made by t and the rays $\overrightarrow{AB}$ and $\overrightarrow{AC}$.

10.1.4. Prove Euclid's Proposition III.32.
Hint: First consider the case in which $\overleftrightarrow{BC}$ is parallel to t. Show that $\triangle ABC$ is an isosceles triangle. In the other case, let O be the point at which $\overleftrightarrow{BC}$ intersects t. It

may be assumed that C is between B and O (explain). Use Euclid's Proposition III.36 and the SAS Similarity Criterion to show that $\triangle OAC \sim \triangle OBA$.

10.2 The radical axis

Now we investigate the locus of points that have the same power with respect to two different circles.

Exercise

*10.2.1. Construct two circles α and β that have different centers. Construct a point P and calculate $p(P, \alpha)$ and $p(P, \beta)$. Move P and determine by experimentation all points P for which $p(P, \alpha) = p(P, \beta)$. Make a conjecture about the shape of the set of such points. Do this for disjoint circles and for circles that intersect. Compare your results.

Definition. The *radical axis* $r(\alpha, \beta)$ of two circles α and β is the locus of points P such that $p(P, \alpha) = p(P, \beta)$; i.e.,

$$r(\alpha, \beta) = \{P \mid p(P, \alpha) = p(P, \beta)\}.$$

The next few exercises will prove that the radical axis of two circles with distinct centers is a line. Exercise 10.2.3 can be viewed as a generalization of the pointwise characterization of perpendicular bisector.

Exercises

10.2.2. Let A and B be two distinct points. Prove that if E and F lie on $\overleftrightarrow{AB}$ and $(EA)^2 - (EB)^2 = (FA)^2 - (FB)^2$, then $E = F$.
Hint: Introduce coordinates on the line so that A corresponds to the real number a, B corresponds to b, E corresponds to e, and F corresponds to f. Translate the equation into one relating a, b, e, and f (for example, $EA = |e - a|$), and show that $e = f$.

10.2.3. Let A and B be two distinct points, let E be a point on $\overleftrightarrow{AB}$, and let ℓ be the line that is perpendicular to $\overleftrightarrow{AB}$ at E. Prove that a point P lies on ℓ if and only if $(PA)^2 - (PB)^2 = (EA)^2 - (EB)^2$.
Hint: The forward implication is an application of the Pythagorean theorem. For the converse, drop a perpendicular from P to $\overleftrightarrow{AB}$ and call the foot F. Use the hypothesis and the previous exercise to show that $E = F$. Conclude that P lies on ℓ.

10.2.4. Let A and B be two distinct points. Prove that for every real number c there exists a unique point X on $\overleftrightarrow{AB}$ such that $(AX)^2 - (BX)^2 = c$.
Hint: Again introduce coordinates on the line so that A corresponds to the real number a, B corresponds to b, and X corresponds to x. Translate the equation into one relating a, b, x, and c. Check that there is a unique solution for x in terms of a, b, and c.

10.2.5. Let α be a circle with center A and radius r. Prove that, for every point P, $p(P, \alpha) = (PA)^2 - r^2$.

10.2.6. Let α be a circle with center A and let β be a circle with center B, $A \neq B$. Combine the last few exercises to prove that $r(\alpha, \beta)$ is a line that is perpendicular to $\overleftrightarrow{AB}$.

Hint: First locate a point E on $\overleftrightarrow{AB}$ such that $(EA)^2 - (EB)^2 = r_\alpha^2 - r_\beta^2$, where r_α and r_β are the radii of α and β, respectively. Then prove that the line that is perpendicular to $\overleftrightarrow{AB}$ at E is the radical axis.

Here is a statement of the theorem that was proved in the last exercise.

Radical Axis Theorem. *If α and β are two circles with different centers, then $r(\alpha, \beta)$ is a line that is perpendicular to the line through the centers of the circles.*

Exercises

10.2.7. Prove that the radical axis of a pair of tangent circles is the common tangent line.

10.2.8. Prove that the radical axis of a pair of circles that intersect at two points is the common secant line.

10.2.9. Suppose α and β are circles that share the same center. What is $r(\alpha, \beta)$ in this case?

10.3 The radical center

Three circles can be taken in pairs to determine three radical axes. The next theorem asserts that they are concurrent, a result that will be useful in the proof of Brianchon's theorem in the next chapter.

Radical Center Theorem. *If α, β, and γ are three circles with distinct centers, then the three radical axes determined by the circles are concurrent (in the extended plane).*

Exercise

10.3.1. Prove the radical center theorem.

Hint: Prove that the radical axes intersect at an ideal point if the centers of the three circles are collinear. If not, prove that any two radical axes intersect and that the third radical axis must pass through the point at which the first two intersect.

Definition. The point of concurrence of the three radical axes of the three circles α, β, and γ is called the *radical center* of the three circles.

11

Applications of the Theorem of Menelaus

The theorem of Menelaus is powerful and has interesting consequences in a variety of situations. This chapter contains a sampling of corollaries.

11.1 Tangent lines and angle bisectors

The first applications are simple results about how tangent lines and angle bisectors intersect the sidelines of the triangle. All the proofs in this section rely on the trigonometric form of Menelaus's theorem.

Exercises

*11.1.1. Construct a triangle and its circumscribed circle. For each vertex of the triangle, construct the line that is tangent to the circumcircle at that point. Mark the point at which the line that is tangent at a vertex intersects the opposite sideline of the triangle. Verify that the three points you have marked are collinear. Under what conditions is one or more of the intersection points an ideal point?

11.1.2. Let $\triangle ABC$ be a triangle. Prove that the lines that are tangent to the circumcircle of $\triangle ABC$ at the vertices of the triangle cut the opposite sidelines at three collinear points (Figure 11.1).
Hint: Use Euclid's Proposition III.32.

*11.1.3. Construct a triangle $\triangle ABC$. At each vertex, construct the line that bisects the two exterior angles at that vertex. Mark the point at which the bisector intersects the opposite sideline. Verify that the three points you have marked are collinear. Under what conditions is one of more of the intersection points an ideal point?

11.1.4. Let $\triangle ABC$ be a triangle. Prove that the external angle bisectors of $\triangle ABC$ meet the opposite sidelines of the triangle in three collinear points.

*11.1.5. Construct a triangle $\triangle ABC$. At each of vertices A and B construct the line that bisects the interior angle at the vertex and mark the point at which the bisector intersects the opposite sideline. At vertex C, construct the line that bisects the exterior

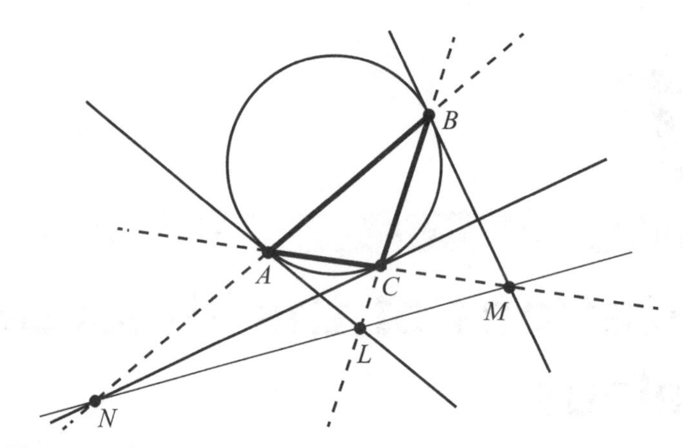

Figure 11.1. The tangent lines cut opposite sidelines at collinear points

angles and mark the point at which the bisector intersects the opposite sideline. Verify that the three points you have marked are collinear. Under what conditions is one or more of the intersection points an ideal point?

11.1.6. Let $\triangle ABC$ be a triangle. Prove that internal angle bisectors at A and B and the external angle bisector at C meet the opposite sidelines of the triangle in three collinear points.

11.2 Desargues' theorem

The theorem in this section is due to Girard Desargues (1591–1661). It has important applications to the theory of perspective drawing.

Definition. Two triangles $\triangle ABC$ and $\triangle A'B'C'$ are said to be *perspective from the point* O if the three lines $\overleftrightarrow{AA'}$, $\overleftrightarrow{BB'}$, and $\overleftrightarrow{CC'}$ joining corresponding vertices are concurrent at O. The point O is called the *perspector* or the *point of perspective* (see Figure 11.2).

Some authors use the term *copolar* to describe triangles that are perspective from a point.

Definition. Two triangles $\triangle ABC$ and $\triangle A'B'C'$ are said to be *perspective from a line* if the three points L, M, and N at which corresponding sidelines $\overleftrightarrow{BC}$ and $\overleftrightarrow{B'C'}$, $\overleftrightarrow{AC}$ and $\overleftrightarrow{A'C'}$, and $\overleftrightarrow{AB}$ and $\overleftrightarrow{A'B'}$ intersect are collinear. The line containing L, M, and N is called the *perspectrix* (see Figure 11.2).

Some authors use the term *coaxial* to describe triangles that are perspective from a line.

Desargues' Theorem. *Two triangles are perspective from a point if and only if they are perspective from a line.*

Exercises

*11.2.1. Make a GeoGebra sketch showing two triangles that are perspective from a point. Verify that they are perspective from a line. What happens when the perspec-

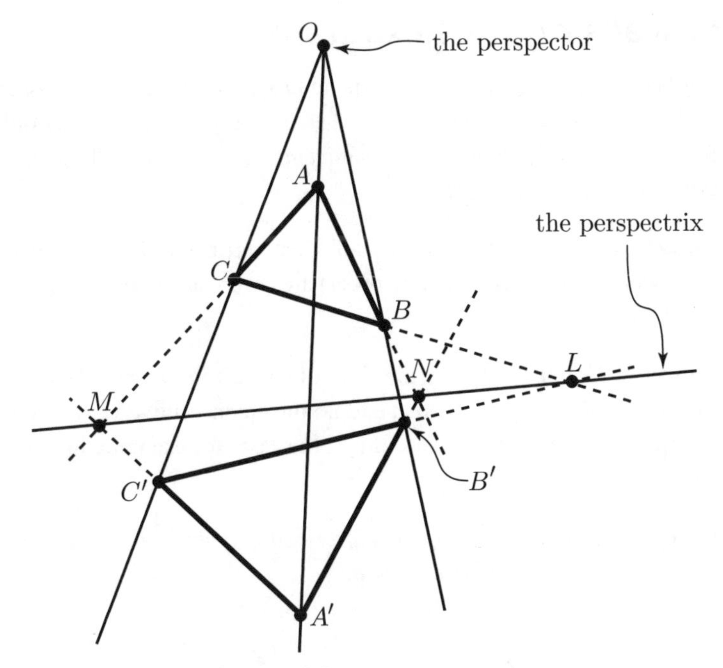

Figure 11.2. Desargues' theorem

tor is an ideal point? What happens when one or more of the pairs of corresponding sidelines of the triangles are parallel? How many pairs of corresponding sides can be parallel?

*11.2.2. Make a GeoGebra sketch showing two triangles that are perspective from a line. Verify that they are perspective from a point.

11.2.3. Prove that if the triangles $\triangle ABC$ and $\triangle A'B'C'$ are perspective from a point, then they are perspective from a line.
Hint: Assume that the two triangles are perspective from O. The points L, M, and N may be defined as the intersections of corresponding sidelines of the two triangles. Explain why it is enough to prove that

$$\frac{\mathbf{AN}}{\mathbf{NB}} \cdot \frac{\mathbf{BL}}{\mathbf{LC}} \cdot \frac{\mathbf{CM}}{\mathbf{MA}} = -1.$$

Apply Menelaus's theorem to the triangle $\triangle OAB$ with collinear Menelaus points N, A', and B'. Apply it in a similar way to $\triangle OAC$ and $\triangle OBC$, and then multiply the resulting equations together.

11.2.4. Prove that if the triangles $\triangle ABC$ and $\triangle A'B'C'$ are perspective from a line, then they are perspective from a point.
Hint: Assume that the points L, M, and N are collinear. Define O to be the point of intersection of $\overleftrightarrow{BB'}$ and $\overleftrightarrow{CC'}$. Explain why it is enough to prove that A, A', and O are collinear. The triangles $\triangle MCC'$ and $\triangle NBB'$ are perspective from the point L, so you can apply the part of the theorem that you have already proved to them.

11.3 Pascal's mystic hexagram

Polygons inscribed in circles have some surprising properties. One of the most interesting was discovered by Blaise Pascal (1623–1662) when he was just sixteen years old. He gave it the colorful Latin title *mysterium hexagrammicum*. For that reason the theorem is still referred to as *Pascal's mystic hexagram.*

Definition. A *hexagon* is a polygon with six distinct vertices $ABCDEF$. It is required that no three consecutive vertices (in cyclic order) are collinear. A hexagon is *inscribed* in the circle γ if all the vertices lie on γ.

Just as with quadrilaterals, we allow the sides of a hexagon to cross. The only requirement on the vertices is that they be distinct and no three consecutive vertices be collinear. Nonconsecutive vertices can be collinear—this is the case, for example, in the theorem of Pappus in §11.5.

Pascal's Mystic Hexagram. *If a hexagon is inscribed in a circle, then the three points at which opposite sidelines intersect are collinear.*

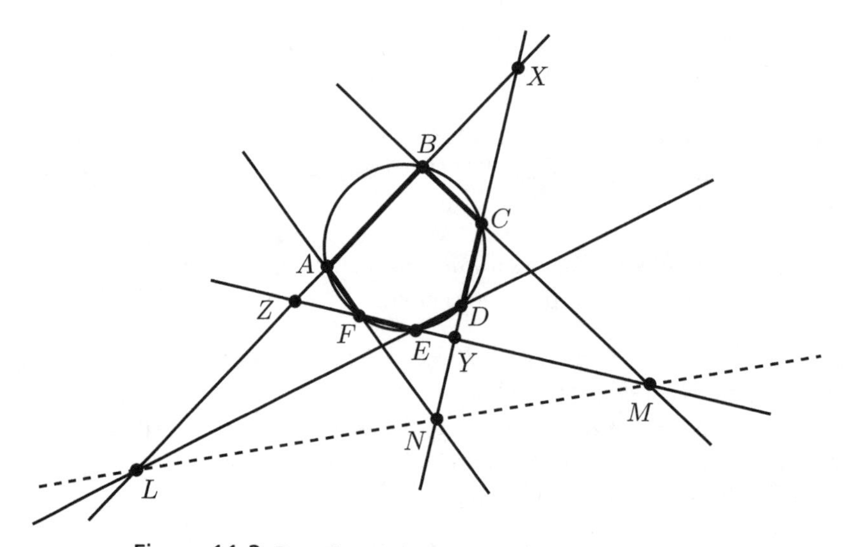

Figure 11.3. Pascal's mystic hexagram for a convex hexagon

The line containing the points at which the opposite sidelines intersect is called the *Pascal line* for the hexagon.

Exercises

11.3.1. Let A, B, C, D, E, and F be distinct points lying on a circle $\mathcal{C}$. How many different hexagons have these six points as vertices?

*11.3.2. Construct a circle and six points A, B, C, D, E, and F, cyclically ordered around the circle. Draw the lines through the sides and mark the points at which the opposite sidelines intersect. Verify that these three points are collinear. Verify that this is the case for both convex and crossed hexagons.

*11.3.3. Draw examples of inscribed hexagons for which zero, one, and three of the points of intersection are ideal points. Is it possible for exactly two of the points to be ideal?

*11.3.4. Draw examples of crossed hexagons inscribed in a circle and verify Pascal's theorem for them.

11.3.5. Explain how the result of Exercise 11.1.2 can be viewed as a limiting case of Pascal's theorem.

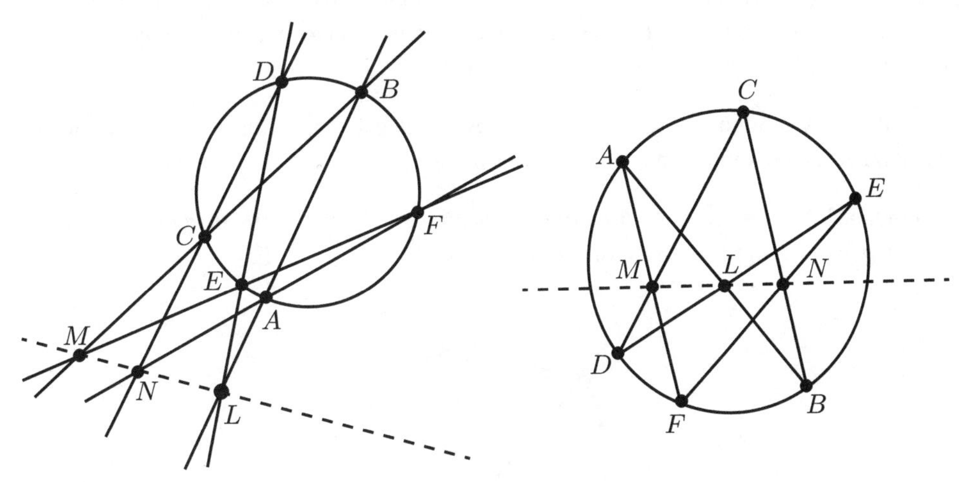

Figure 11.4. Pascal's theorem for two crossed hexagons

Notation. Let γ be a circle, and let A, B, C, D, E, and F be six points on γ, cyclically ordered around the circle. Define six additional points:

L is the point at which $\overleftrightarrow{AB}$ meets $\overleftrightarrow{DE}$
M is the point at which $\overleftrightarrow{BC}$ meets $\overleftrightarrow{EF}$
N is the point at which $\overleftrightarrow{CD}$ meets $\overleftrightarrow{AF}$
X is the point at which $\overleftrightarrow{AB}$ meets $\overleftrightarrow{CD}$
Y is the point at which $\overleftrightarrow{EF}$ meets $\overleftrightarrow{CD}$
Z is the point at which $\overleftrightarrow{AB}$ meets $\overleftrightarrow{EF}$.

To keep the proof of Pascal's theorem relatively simple, we will consider only the case in which X, Y, and Z are ordinary points.

Exercises

*11.3.6. Use GeoGebra to make a sketch showing the circle γ and the twelve points listed above. Observe that L, M, and N are Menelaus points for $\triangle XYZ$.

11.3.7. Prove Pascal's theorem if all three of the points X, Y, and Z are ordinary points. Hint: The proof is accomplished by using Menelaus's theorem to show that L, M, and N are collinear Menelaus points for the triangle $\triangle XYZ$. Apply Menelaus's theorem three times to $\triangle XYZ$ with collinear Menelaus points $\{B, C, M\}$, $\{A, F, N\}$, and $\{D, E, L\}$. Multiply the resulting equations together and apply Euclid's Proposition III.36 to reach the desired conclusion.

11.4 Brianchon's theorem

As mentioned earlier, Brianchon's theorem is dual to Pascal's theorem. It illustrates a new aspect of duality: a point on circle is dual to a line that is tangent to the circle. The theorem is included in this chapter on applications of Menelaus's theorem because it is such a beautiful example of duality, not because it is an application of the theorem of Menelaus. Instead the proof is based on the radical center theorem. Brianchon's theorem is named for its discoverer, Charles Julien Brianchon (1783–1864).

Definition. A hexagon is said to be *circumscribed about a circle* if each side of the hexagon is tangent to the circle at an interior point.

Brianchon's Theorem. *If a hexagon is circumscribed about a circle, then the lines determined by pairs of opposite vertices are concurrent.*

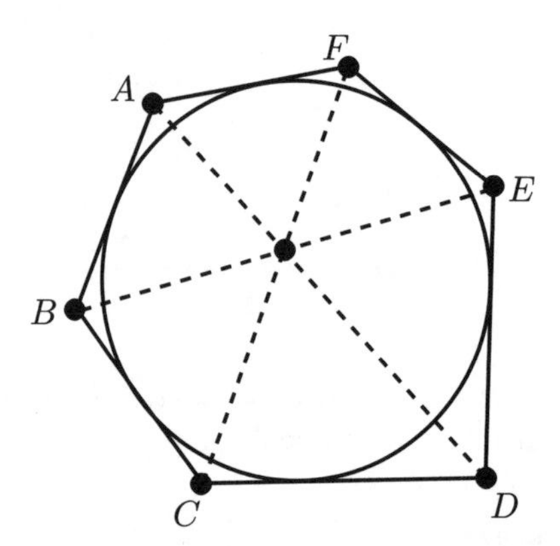

Figure 11.5. Brianchon's theorem for a convex hexagon

Exercises

*11.4.1. Make a GeoGebra worksheet that can be used to verify and illustrate Brianchon's theorem for convex hexagons.

*11.4.2. Find examples of crossed hexagons that satisfy the hypotheses of Brianchon's theorem. Verify that the theorem holds for them as well.

The proof of Brianchon's theorem relies on a construction that is fairly easy to reproduce with GeoGebra, but which few people would think of for themselves. We will describe the construction first, and then look at the proof itself. As you read the next two paragraphs you should make a GeoGebra sketch that reproduces Figure 11.7.

Let $\mathcal{C}$ be a circle and let $ABCDEF$ be a hexagon that is circumscribed about $\mathcal{C}$. This means that all the vertices of $ABCDEF$ lie outside $\mathcal{C}$ and each side of $ABCDEF$ is tangent to $\mathcal{C}$ at an interior point of that side. Let P, Q, R, S, T, and U be the points of tangency, labeled as in Figure 11.7.

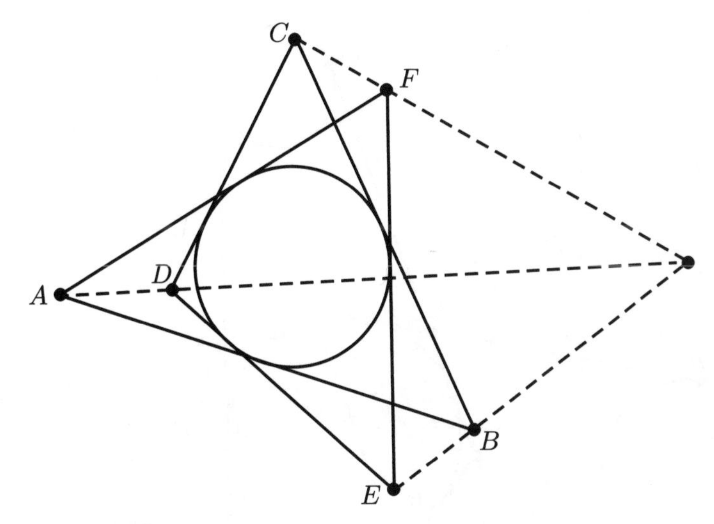

Figure 11.6. Brianchon's theorem for a crossed hexagon

Construct points P' on $\overrightarrow{FA}$, Q' on $\overrightarrow{BA}$, R' on $\overrightarrow{BC}$, S' on $\overrightarrow{DC}$, T' on $\overrightarrow{DE}$, and U' on $\overrightarrow{FE}$ such that $PP' = QQ' = RR' = SS' = TT' = UU'$. (The precise value of these distances is not important, provided all six are equal.) Construct three circles α, β, and γ such that α is tangent to $\overleftrightarrow{AF}$ at P' and tangent to $\overleftrightarrow{CD}$ at S', β is tangent to $\overleftrightarrow{BC}$ at R' and tangent to $\overleftrightarrow{FE}$ at U', and γ is tangent to $\overleftrightarrow{AB}$ at Q' and tangent to $\overleftrightarrow{DE}$ at T'. The circles exist by Exercise 11.4.3, below.

Exercises

11.4.3. Let $\mathcal{C}$ be a circle and let t and s be two lines that are tangent to $\mathcal{C}$ at points P and S, respectively. If P' is a point on t and $S' \neq P'$ is a point on s such that P' and S' are on the same side of $\overleftrightarrow{PS}$ and $PP' = SS'$, then there exists a circle α that is tangent to t and s at P' and S'.
Hint: The circles $\mathcal{C}$ and α in Figure 11.7 illustrate the exercise. First prove the theorem if $\overleftrightarrow{PP'} \parallel \overleftrightarrow{SS'}$ and then use similar triangles to prove the other case.

11.4.4. Prove that $FP' = FU'$, $CS' = CR'$, $BQ' = BR'$, $ET' = EU'$, $AQ' = AP'$, and $DT' = DS'$ in Figure 11.7.

11.4.5. Prove that $r(\alpha, \beta) = \overleftrightarrow{CF}$, $r(\beta, \gamma) = \overleftrightarrow{BE}$, and $r(\alpha, \gamma) = \overleftrightarrow{AD}$.

11.4.6. Use the preceding exercise and the radical center theorem to prove Brianchon's theorem.

11.5 Pappus's theorem

The next theorem was discovered by Pappus of Alexandria. Pappus, who lived from approximately AD 290 until about 350, was one of the last great geometers of antiquity. Much later his theorem became an important result in the foundations of projective geometry. Both the statement and proof of Pappus's theorem are reminiscent of Pascal's theorem.

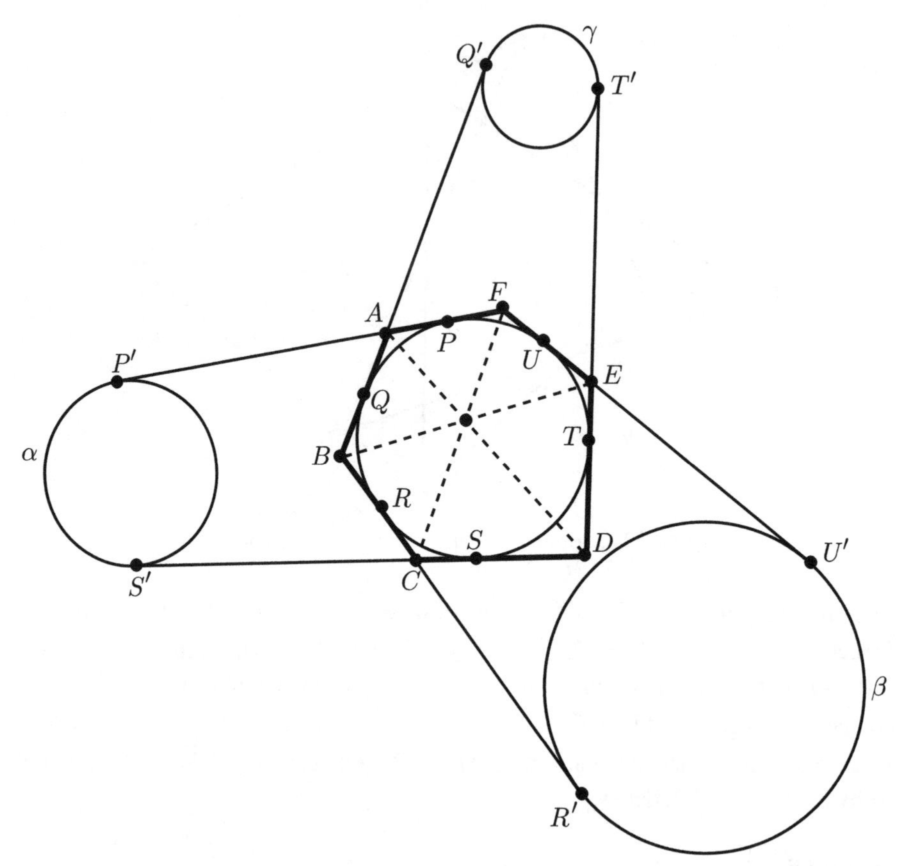

Figure 11.7. Construction for the proof of Brianchon's theorem

There is a hexagon involved, but the vertices of the hexagon lie on two lines rather than on a circle. The hexagon is necessarily a crossed hexagon.

Pappus's Theorem. *Let A, B, C, D, E, and F be six points. Define L to be the point at which $\overleftrightarrow{AB}$ meets $\overleftrightarrow{DE}$, define M to be the point at which $\overleftrightarrow{BC}$ meets $\overleftrightarrow{EF}$, and define N to be the point at which $\overleftrightarrow{CD}$ meets $\overleftrightarrow{AF}$. If A, C, E lie on one line and B, D, F lie on another line, then the points L, M, and N are collinear.*

Exercise

*11.5.1. Make a GeoGebra worksheet that can be used to verify and illustrate Pappus's theorem.

Before we begin the proof of the theorem, let us establish some notation (see Figure 11.8).

Notation. Start with six points A, B, C, D, E, and F. Assume A, C, and E lie on a line ℓ and that B, D, and F lie on a line m. Define six additional points:

L is the point at which $\overleftrightarrow{AB}$ meets $\overleftrightarrow{DE}$

M is the point at which $\overleftrightarrow{BC}$ meets $\overleftrightarrow{EF}$

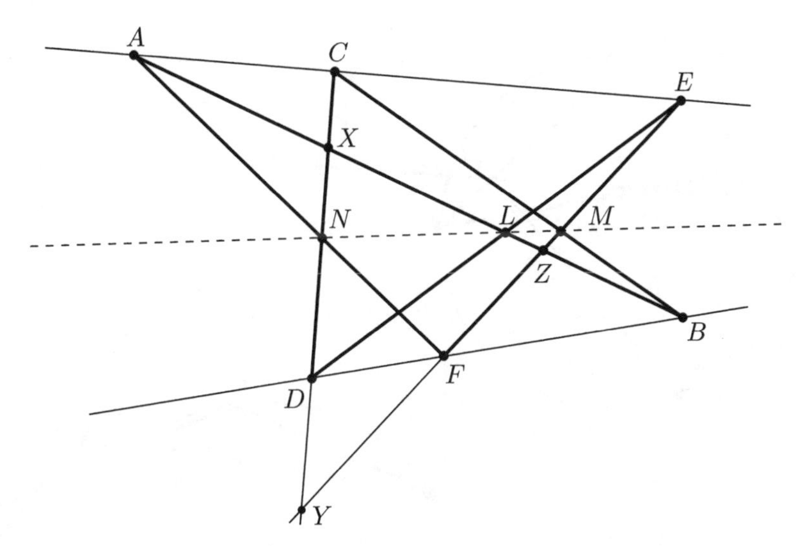

Figure 11.8. Pappus's theorem

N is the point at which $\overleftrightarrow{CD}$ meets $\overleftrightarrow{AF}$
X is the point at which $\overleftrightarrow{AB}$ meets $\overleftrightarrow{CD}$
Y is the point at which $\overleftrightarrow{EF}$ meets $\overleftrightarrow{CD}$
Z is the point at which $\overleftrightarrow{AB}$ meets $\overleftrightarrow{EF}$.

To keep the proof relatively simple, we will prove Pappus's theorem only when the three points X, Y, and Z are ordinary points.

Exercises

*11.5.2. Use GeoGebra to make a sketch showing the ℓ and m along with the twelve points listed above. Observe that L, M, and N are Menelaus points for $\triangle XYZ$.

11.5.3. Prove Pappus's theorem if X, Y, and Z are ordinary points.
Hint: The proof is accomplished by using Menelaus's theorem to show that L, M, and N are collinear Menelaus points for the triangle $\triangle XYZ$. Apply Menelaus's theorem five times to $\triangle XYZ$ with collinear Menelaus points $\{B, C, M\}$, $\{A, F, N\}$, $\{D, E, L\}$, $\{A, C, E\}$, and $\{B, D, F\}$. Multiply the first three equations, divide by the product of the last two, and cancel lots of terms.

11.6 Simson's theorem

The theorem in this section is usually ascribed to the Scottish mathematician Robert Simson (1687–1768), but there is no record that he ever published it. The theorem was published in 1799 by William Wallace (1768–1843). The result was discussed earlier in the context of pedal triangles.

Simson's Theorem. *A point P is on the circumscribed circle of triangle $\triangle ABC$ if and only if the feet of the perpendiculars from P to the sidelines of $\triangle ABC$ are collinear.*

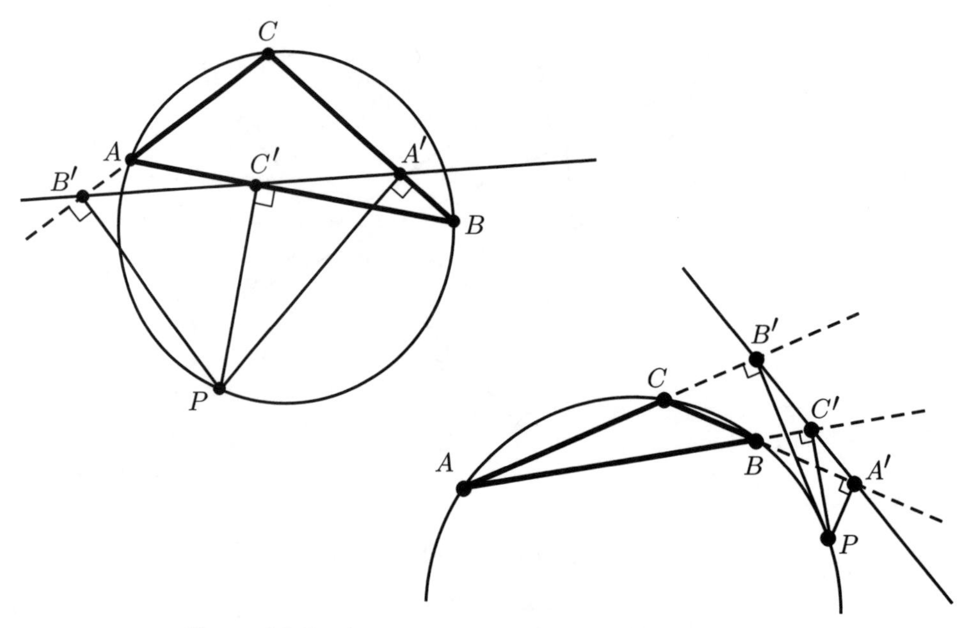

Figure 11.9. Two possible configurations in Simson's theorem

Definition. A line that contains the feet of three perpendiculars from a point P to the triangle $\triangle ABC$ is called a *Simson line* for $\triangle ABC$. The point P is the *pole* of the Simson line.

Notation. As in Chapter 5, we will consistently use A' to denote the foot of the perpendicular to the sideline $\overleftrightarrow{BC}$, B' to denote the foot of the perpendicular to $\overleftrightarrow{AC}$, and C' to denote the foot of the perpendicular to $\overleftrightarrow{AB}$ (see Figure 11.9).

One way to state Simson's theorem is to say that the pedal triangle determined by P degenerates to a line segment exactly when the point P is on the circumcircle. For that reason a Simson line is often called a *pedal line*. Before tackling the proof of Simson's theorem we will use GeoGebra to explore several interesting properties of Simson lines.

Exercises

*11.6.1. Construct a triangle $\triangle ABC$ and its circumcircle. Choose a point P and construct the feet of the three perpendiculars from P to the sidelines of the triangle. Put a line through two of the feet and verify that the third foot is on that line if and only if P is on the circumcircle.
Hint: In order to make sure that you consider all possible shapes for $\triangle ABC$, it is best to construct the circumcircle first and then construct A, B, and C to be three movable points on the circle.

*11.6.2. Find triangles $\triangle ABC$ and points P on the circumcircle of $\triangle ABC$ such that the Simson line with pole P intersects $\triangle ABC$ in exactly two points. Now find examples for which the Simson line is disjoint from the triangle. Make observations about the shape of the triangles and the location of P for which the latter possibility occurs.

*11.6.3. Let Q be the point at which the altitude through A meets the circumcircle. Verify that the Simson line with pole Q is parallel to the tangent to the circle at A.

*11.6.4. Let P and Q be two points on the circumcircle of $\triangle ABC$. Verify that the measure of the angle between the Simson lines with poles P and Q is half the measure of $\angle POQ$ (where O is the circumcenter).

*11.6.5. Verify that if $\overline{PQ}$ is a diameter for the circumcircle, then the Simson lines with poles P and Q intersect at a point on the nine-point circle.

*11.6.6. Let H be the orthocenter of $\triangle ABC$. Verify that if P is a point on the circumcircle, then the midpoint of $\overline{HP}$ lies on the Simson line with pole P.

The next four exercises outline a proof of the forward implication in Simson's theorem. We will assume that P is a point on the circumcircle of $\triangle ABC$ and prove that the feet of the perpendiculars from P to the sidelines of $\triangle ABC$ are collinear. In case P is at one of the vertices, the feet of two of the perpendiculars coincide with P and the three feet are obviously collinear. Thus we need only consider the case in which P is distinct from the vertices. Relabel the vertices of the triangle so that P is between A and B in cyclic order on the circumcircle (see Figure 11.9). Then $\square APBC$ is a convex cyclic quadrilateral.

Exercises

11.6.7. Prove that $\angle PBA \cong \angle PCA$ and then use similar triangles to conclude that $\dfrac{\mathbf{CB'}}{\mathbf{BC'}} = \dfrac{\mathbf{PB'}}{\mathbf{PC'}}$.

11.6.8. Prove that $\angle PAB \cong \angle PCB$ and then use similar triangles to conclude that $\dfrac{\mathbf{AC'}}{\mathbf{CA'}} = \dfrac{\mathbf{PC'}}{\mathbf{PA'}}$.

11.6.9. Use Euclid's Proposition III.22 to prove that $\angle PBC \cong \angle PAB'$ and then use similar triangles to conclude that $\dfrac{\mathbf{BA'}}{\mathbf{AB'}} = \dfrac{\mathbf{PA'}}{\mathbf{PB'}}$.

11.6.10. Use the theorem of Menelaus to prove that if P lies on the circumcircle of $\triangle ABC$, then A', B', and C' are collinear.

The proof of the converse to Simson's theorem is a bit complicated, so we will not supply complete details. Instead we will make some simplifying assumptions and base our proof on them. It can be shown that the assumptions are always satisfied [3, page 40], but we will omit that proof. A different proof of the converse of Simson's theorem may be found in [9, §3A].

In the following exercises, $\triangle ABC$ is a triangle, P is a point in the interior of $\angle ACB$, and $\square APBC$ is a convex quadrilateral. Assume A', B', and C' are collinear. We will prove that, given those assumptions, P lies on the circumcircle of $\triangle ABC$.

Exercises

11.6.11. Prove that $\angle B'PA$ and $\angle B'C'A$ are either congruent or supplements.
Hint: First show that the four points B', P, C', and A lie on a circle.

11.6.12. Prove that $\angle BPA'$ and $\angle BC'A'$ are either congruent or supplements.
Hint: This time show that P, B, A', and C' lie on a circle.

11.6.13. Assume, in addition to the hypotheses above, that C' lies on $\overline{AB}$. Use Euclid's Proposition III.22 and its converse to prove that P lies on the circumcircle of $\triangle ABC$.

11.6.14. Assume that B is between A and C'. Use Euclid's Proposition III.22 and its converse to prove that P lies on the circumcircle of $\triangle ABC$.

The remaining case (A is between B and C') follows by a similar argument.

We conclude our treatment of Simson's theorem with a beautifully simple application. Both directions of Simson's theorem will be required to prove that four special circles meet in a single point. First we need a definition.

Definition. Four distinct lines in the plane are said to be in *general position* if no two of them are parallel and no three of them are concurent.

The four lines determine six points of intersection and four triangles (Figure 11.10). We will prove that the four circumcircles of the triangles have a common point of intersection.

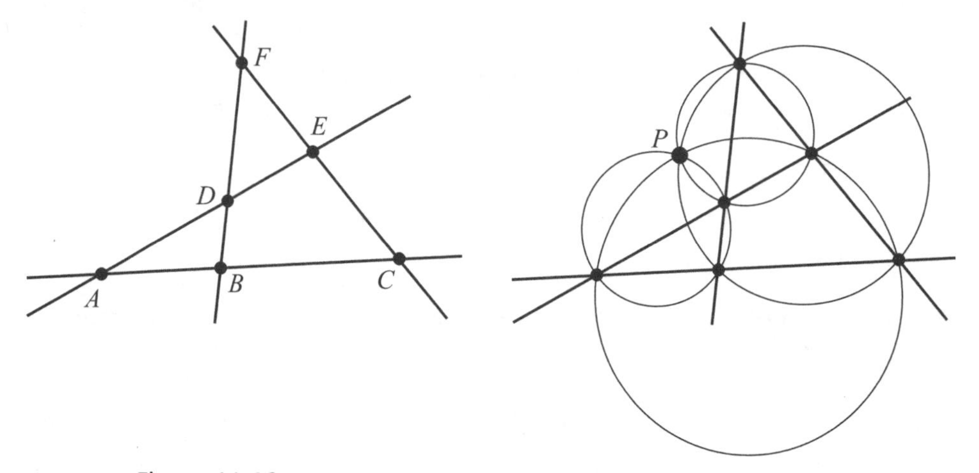

Figure 11.10. Four lines in general position determine four circumcircles

Exercises

*11.6.15. Construct four lines in general position, the four triangles they determine, and the four circumcircles for the four triangles. Verify that the four circumcircles have a common point of intersection.

11.6.16. Prove the result you verified in the previous exercise.

11.7 Ptolemy's theorem

Ptolemy's theorem is a useful result regarding cyclic quadrilaterals that is attributed to Claudius Ptolemaeus of Alexandria (85–165). It asserts that the product of the lengths of the two diagonals of a cyclic quadrilateral is equal to the sum of the products of the lengths of the opposite sides. Here is a precise statement:

Ptolemy's Theorem. *If $\square ABCD$ is a cyclic quadrilateral, then*

$$AB \cdot CD + BC \cdot AD = AC \cdot BD.$$

Exercises

*11.7.1. Make a GeoGebra worksheet that illustrates and verifies Ptolemy's theorem. Part of the definition of cyclic quadrilateral requires that the quadrilateral be convex; is the theorem valid for crossed quadrilaterals whose vertices lie on a circle?

*11.7.2. Construct a circle γ and three movable points A, B, and C on γ. Construct an additional point P that lies on the arc of γ from A to C that does not contain B. As in the previous section, denote the feet of the perpendiculars from P to the sidelines by $\triangle ABC$ by A', B', and C'. Verify that B' is always between A' and C', whether the Simson line intersects $\triangle ABC$ or not.

11.7.3. Prove the result you verified in the last exercise.

11.7.4. Prove Ptolemy's theorem.
Hint: Consider the Simson line for $\triangle ABC$ with pole D. By Exercise 11.7.3, B' is between A' and C', so $A'B' + B'C' = A'C'$. Apply Exercise 5.4.5 to reach the desired conclusion.

11.8 The butterfly theorem

Our final application of the theorem of Menelaus is known as the butterfly theorem. The reason for the name is evident from Figure 11.11.

Butterfly Theorem. *Assume γ is a circle, $\overline{PQ}$ is a chord of γ, and M is the midpoint of $\overline{PQ}$. Let $\overline{AB}$ and $\overline{CD}$ be two chords of γ such that A and C are on the same side of $\overleftrightarrow{PQ}$ and both $\overline{AB}$ and $\overline{CD}$ intersect $\overline{PQ}$ at M. If $\overline{AD}$ intersects $\overline{PQ}$ at X and $\overline{BC}$ intersects $\overline{PQ}$ at Y, then M is the midpoint of $\overline{XY}$.*

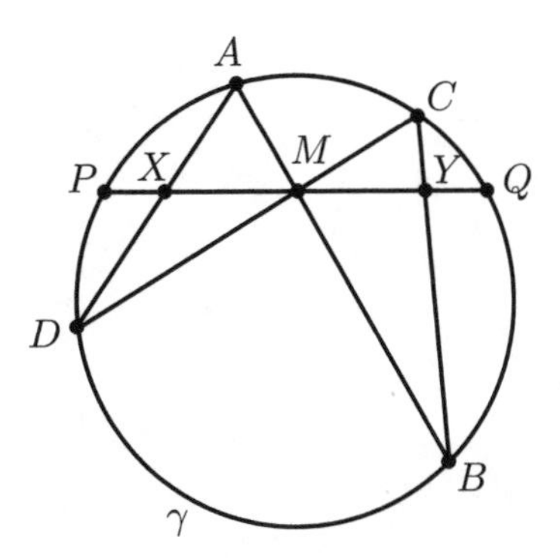

Figure 11.11. The butterfly theorem

Exercises

*11.8.1. Make a GeoGebra worksheet that illustrates and verifies the butterfly theorem.

*11.8.2. Verify that the butterfly theorem remains true even when the assumption that A and C are on the same side of $\overleftrightarrow{PQ}$ is dropped. In that case the segment $\overline{PQ}$ must be extended to the line $\overleftrightarrow{PQ}$ and the points X and Y lie outside γ. Can you find a butterfly in this diagram? While the result is true in this generality, we will only prove the theorem as stated above.

11.8.3. Prove the butterfly theorem in the special case in which $\overleftrightarrow{AD} \parallel \overleftrightarrow{BC}$.
Hint: Prove that M must equal the center of γ (in this case). Then use ASA to prove that $\triangle DMX \cong \triangle CMY$.

11.8.4. Prove the following simple result from elementary algebra; it is needed in the proof of the remaining case of the butterfly theorem: *If x, y, and m are three positive numbers such that*

$$\frac{x^2(m^2 - y^2)}{y^2(m^2 - x^2)} = 1,$$

then $x = y$.

11.8.5. Prove the butterfly theorem in case $\overleftrightarrow{AD}$ intersects $\overleftrightarrow{BC}$.
Hint: Let R be the point at which $\overleftrightarrow{AD}$ and $\overleftrightarrow{BC}$ intersect. To simplify the notation, let $x = MX$, $y = MY$, and $m = PM$. Apply Menelaus's theorem twice to $\triangle RXY$, first with collinear Menelaus points $\{A, M, B\}$ and then with collinear Menelaus points $\{C, M, D\}$. Multiply the resulting equations. Apply Euclid's Proposition III.36 three times to conclude that $RA \cdot RD = RB \cdot RC$, $XA \cdot XD = XP \cdot XQ$, and $YB \cdot YC = YP \cdot YQ$. All of that should result in the equation

$$\frac{x^2(m^2 - y^2)}{y^2(m^2 - x^2)} = 1.$$

Use Exercise 11.8.4 to complete the proof.

12

Additional Topics in Triangle Geometry

This chapter examines a number of loosely connected topics regarding the geometry of the triangle. We will explore the statements of the theorems using GeoGebra, but will not prove them.

12.1 Napoleon's theorem and the Napoleon point

The theorem in this section is commonly attributed to the French emperor Napoleon Bonaparte (1769-1821). Napoleon was an amateur mathematician, with a particular interest in geometry, who took pride in his mathematical talents. Thus this attribution may be based at least partially on historical fact. On the other hand, Coxeter and Greitzer [3, page 63] make the following comment regarding the possibility that the theorem might in fact be due to Napoleon: "...the possibility of his knowing enough geometry for this feat is as questionable as the possibility of his knowing enough English to compose the famous palindrome ABLE WAS I ERE I SAW ELBA."

Exercises

*12.1.1. Construct a triangle $\triangle ABC$ and use the tool you made in Exercise 3.1.1 to construct an equilateral triangle on each side of $\triangle ABC$. Make sure that the new triangles are on the outside of the original triangle. Label vertices so that the three new triangles are $\triangle A'BC$, $\triangle AB'C$, and $\triangle ABC'$. Construct the centroids of triangles $\triangle A'BC$, $\triangle AB'C$, and $\triangle ABC'$ and label them U, V, and W, respectively. Verify the following.

(a) $\triangle UVW$ is equilateral.

(b) Lines $\overleftrightarrow{AU}$, $\overleftrightarrow{BV}$, and $\overleftrightarrow{CW}$ are concurrent.

*12.1.2. Do the same construction as in the previous exercise, but this time construct the equilateral triangles so that they are oriented towards the inside of $\triangle ABC$. (The equilateral triangles will overlap and may even stick out of $\triangle ABC$.) Do the same two conclusions hold?

Definition. The triangle $\triangle UVW$ in Exercise 12.1.1 is called the *Napoleon triangle* associated with $\triangle ABC$. *Napoleon's theorem* is the assertion that the Napoleon triangle is always equilateral, regardless of the shape of $\triangle ABC$. The point at which lines $\overleftrightarrow{AU}$, $\overleftrightarrow{BV}$, and $\overleftrightarrow{CW}$ concur is called the *Napoleon point* of $\triangle ABC$. It is another new triangle center.

12.2 The Torricelli point

There is still another triangle center that is closely related to the Napoleon point. This new point is called the *Torricelli point* of the triangle. Do not confuse it with the Napoleon point. The point is named for the Italian mathematician Evangelista Torricelli (1608–1647).

Exercises

*12.2.1. Construct a triangle $\triangle ABC$ and construct external equilateral triangles on the three sides of the triangle. Label the vertices as in Exercise 12.1.1. Construct the lines $\overleftrightarrow{AA'}$, $\overleftrightarrow{BB'}$, and $\overleftrightarrow{CC'}$. Note that they are concurrent. The point at which they concur is called the *Torricelli point* of $\triangle ABC$.

*12.2.2. For which triangles is the Torricelli point inside the triangle and for which is it outside?

*12.2.3. For which triangles is the Torricelli point equal to the Napoleon point?

*12.2.4. Construct the circumcircles for the triangles $\triangle A'BC$, $\triangle AB'C$, and $\triangle ABC'$. Observe that all three of these circles pass through the Torricelli point. Thus an alternative way to define the Torricelli point would be to say that it is the point at which the three circumcircles of the external equilateral triangles intersect.

12.3 van Aubel's theorem

There is also a theorem for quadrilaterals that is closely related to Napoleon's theorem for triangles. The theorem for quadrilaterals is known as *van Aubel's theorem.* The *center of a square* is the point at which the diagonals intersect. The center of a square is obviously equidistant from the vertices.

Exercises

*12.3.1. Construct a convex quadrilateral $\square ABCD$. Construct a square on each side of the quadrilateral. Make sure the four squares are all on the outside of the quadrilateral. Construct the centers of the four squares and the two segments joining the centers of the squares based on opposite sides of $\square ABCD$. Measure the lengths of these segments. What do you observe? Measure the angles between the two segments. What do you observe?

*12.3.2. Now try the same thing with the squares constructed towards the inside of the quadrilateral. Is the result still true?

van Aubel's Theorem. *If external squares are constructed on the sides of a convex quadrilateral, then the segments joining the centers of the squares on opposite sides are perpendicular and have equal length.*

12.4 Miquel's theorem and Miquel points

The theorem in this section is attributed to the nineteenth century French mathematician Auguste Miquel. We have proved many theorems regarding concurrent lines; by contrast, this theorem gives a condition under which three circles are concurrent.

Exercises

*12.4.1. Construct a triangle $\triangle ABC$ and movable points D, E, and F on the sides $\overline{BC}$, $\overline{AC}$, and $\overline{AB}$, respectively. Now construct the circumcircles for the three triangles $\triangle AEF$, $\triangle BDF$, and $\triangle CDE$. The three circles are concurrent regardless of the shape of $\triangle ABC$ and of how the points D, E and F are chosen. Find examples for which the point of concurrence is inside the triangle and examples for which it is outside.

*12.4.2. What happens if D, E, and F are chosen to lie on the sidelines of $\triangle ABC$ rather than the sides?

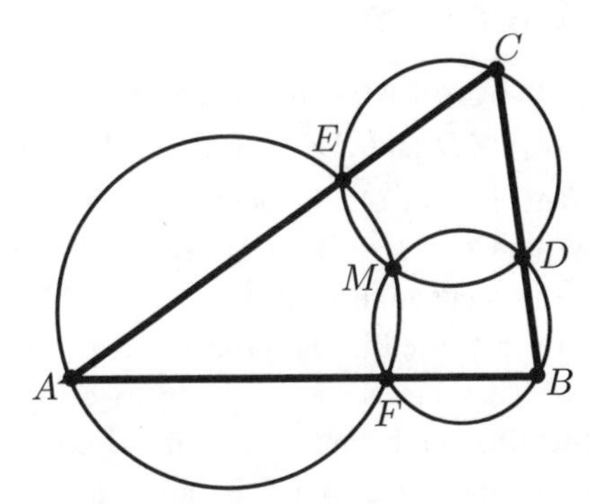

Figure 12.1. Miquel's theorem

The assertion that the three circles in Exercise 12.4.1 are concurrent is known as *Miquel's theorem*. The point of concurrence is called a *Miquel point* for the triangle.

12.5 The Fermat point

Our next triangle center is named for Pierre de Fermat (1601–1665). Apparently Fermat asked Torricelli whether he could locate the point in a triangle at which the sum of the distances to the vertices is smallest. Torricelli solved Fermat's problem and, as we will see, the point he found is closely related to the point that we have called the Torricelli point.

Definition. The *Fermat point* of a triangle is the point F for which the sum of the distances from F to the vertices is as small as possible; i.e., the Fermat point of $\triangle ABC$ is the point F such that $FA + FB + FC$ is minimized.

In the next exercises you will see that for most triangles the Fermat point is the point that makes equal angles with each of the vertices. For such triangles the Fermat point is the same as the Torricelli point.

Exercises

*12.5.1. Construct a triangle $\triangle ABC$ such that all its angles measure less than 120°. Construct a point P in the interior of the triangle and calculate $PA + PB + PC$. Move P around until you locate the Fermat point. Once you have found the point P for which $PA + PB + PC$ is minimized, measure the angles $\angle APB$, $\angle BPC$, and $\angle CPA$. Observe that the sum of the distances is minimized at the same point where the three angles are congruent.

*12.5.2. Now construct a triangle $\triangle ABC$ in which angle $\angle BAC$ has measure greater than 120°. Where is the Fermat point for this triangle located?

*12.5.3. Construct a triangle $\triangle ABC$ and its Torricelli point. Verify that the Torricelli point T and the Fermat point F are the same in case all angles in the triangle have measure less than 120°. Find examples of triangles for which the Torricelli point and the Fermat point are different.

12.6 Morley's theorem

Our final theorem requires that we trisect the angles of a triangle. To *trisect* an angle $\angle BAC$ means to find two rays $\overrightarrow{AD}$ and $\overrightarrow{AE}$ such that $\overrightarrow{AD}$ is between $\overrightarrow{AB}$ and $\overrightarrow{AC}$ and $\mu(\angle BAD) = (1/3)\mu(\angle BAC)$ while $\overrightarrow{AE}$ is between $\overrightarrow{AD}$ and $\overrightarrow{AC}$ and $\mu(\angle DAE) = (1/3)\mu(\angle BAC)$. It follows that $\mu(\angle EAC) = (1/3)\mu(\angle BAC)$ as well, so the original angle is divided into three congruent angles.

The ancient Greeks were interested in the problem of trisecting an angle using only a straightedge and compass. They never succeeded in trisecting an arbitrary angle, and it has been known since the nineteenth century that it is impossible to do so using only the Euclidean compass and straightedge. But it is quite easy to trisect angles using the measurement and calculation capabilities of GeoGebra.

Exercises

*12.6.1. Make a tool that trisects an angle. The tool should accept three points A, B, and C as input objects and should produce as outputs the two rays $\overrightarrow{AB}$ and $\overrightarrow{AC}$ together with the two additional rays in the interior of $\angle BAC$ that trisect the angle.
Hint: Measure the angle $\angle BAC$ and calculate $(1/3)\mu(\angle BAC)$. Rotate $\overrightarrow{AB}$ through $(1/3)\mu(\angle BAC)$ to construct the first trisecting ray and then rotate again through the same angle to produce the second trisecting ray.

*12.6.2. Construct a triangle $\triangle ABC$. Construct both angle trisectors for each of the interior angles of $\triangle ABC$. The rays intersect in twelve points in the interior of $\triangle ABC$. Label the point at which the rays through B and C that are closest to $\overline{BC}$ intersect as A'. Similarly, label the intersection of the two trisectors closest to $\overline{AC}$ as B' and label the intersection of the two trisectors closest to $\overline{AB}$ as C'. The triangle $\triangle A'B'C'$ is called the *Morley triangle* for the triangle $\triangle ABC$. (See Figure 12.2.)

*12.6.3. Hide all the angle trisectors and concentrate on the Morley triangle $\triangle A'B'C'$. Measure all the sides and the angles of $\triangle A'B'C'$ and verify that it is equilateral.

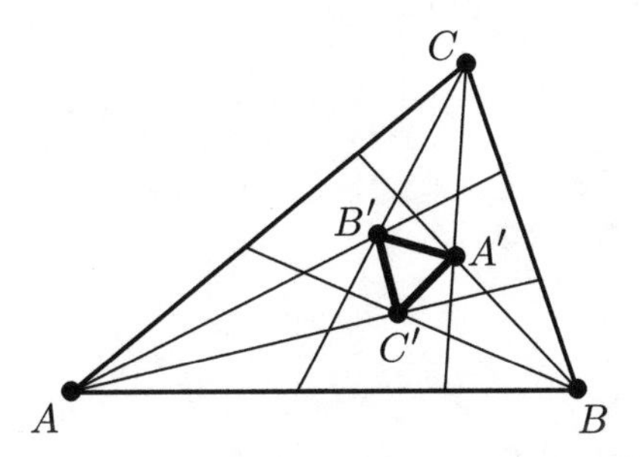

Figure 12.2. $\triangle A'B'C'$ is the Morley triangle for $\triangle ABC$

The theorem asserting that the Morley triangle is equilateral regardless of the shape of the original triangle is known as *Morley's theorem.* It was discovered around 1904 by the American mathematician Frank Morley (1869–1937). It is the most recently discovered theorem we have studied.

13

Inversions in Circles

We now investigate a class of transformations of the Euclidean plane called inversions in circles. The study of inversions is a standard part of college geometry courses, so we will leave most of the details of the proofs to those courses and will not attempt to give a complete treatment. Instead we will concentrate on the construction of the GeoGebra tools that will be needed in the study of the Poincaré disk model of hyperbolic geometry in the next chapter. We will state the basic definitions and theorems, but will not prove the theorems. Proofs may be found in many sources; we will use §10.7 of [11] as our basic reference. In the exercises you will be asked to use the theorems to verify that certain constructions work and then to make GeoGebra tools based on those constructions.

13.1 Inverting points

Let us begin with the definition of inversion.

Definition. Let $\mathcal{C} = \mathcal{C}(O, r)$ be a circle. The *inverse of* P *in* $\mathcal{C}$ is the point $P' = I_{O,r}(P)$ on $\overrightarrow{OP}$ such that $(OP)(OP') = r^2$.

Observe that $I_{O,r}(P)$ is defined for any $P \neq O$, but that the definition breaks down when $P = O$. As P approaches O, the point $I_{O,r}(P)$ will move farther and farther from O. To define $I_{O,r}$ at O, we extend the plane by adding a *point at infinity*, denoted by the symbol ∞. The plane together with this one additional point at infinity is called the *inversive plane*. We extend $I_{O,r}$ to the inversive plane by defining $I_{O,r}(O) = \infty$ and $I_{O,r}(\infty) = O$.

The inversive plane is quite different from the extended Euclidean plane, which was introduced in Chapter 8. In particular, the extended Euclidean plane includes an infinite number of distinct ideal points, while the inversive plane includes just one point at infinity. The inversion $I_{O,r}$ is a transformation of the inversive plane. (In geometry a *transformation* is defined to be a function of a space to itself that is both one-to-one and onto.) There is just one inversive plane on which every inversion is defined (not a different inversive plane for each inversion).

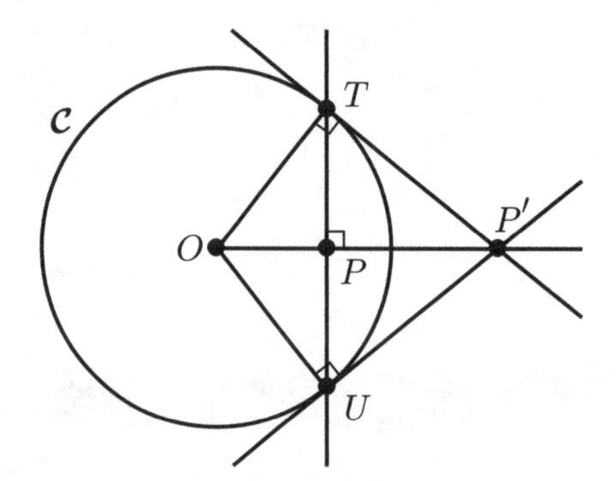

Figure 13.1. Construction of P' if P is inside $\mathcal{C}$

Construction. Let P be a point inside $\mathcal{C}$. Construct the perpendicular to $\overleftrightarrow{OP}$ at P, define T and U to be the points at which the perpendicular intersects $\mathcal{C}$, and then construct P' as shown in Figure 13.1.

Exercises

*13.1.1. Carry out the construction in GeoGebra. Measure OP, OP', and OT and verify that $(OP)(OP') = (OT)^2$.

13.1.2. Prove that the point P' constructed above is the inverse of P.

*13.1.3. Use the construction to make a tool that finds $P' = I_{O,r}(P)$ for each point P inside $\mathcal{C}$. Your tool should accept three points as input objects (the center O, a point on $\mathcal{C}$, and the point P) and it should return the point P' as its output.

Construction. Let P be a point outside $\mathcal{C}$. Find the midpoint M of $\overline{OP}$, construct a circle α with center M and radius MP, and let T and U be the two points of $\alpha \cap \mathcal{C}$. Then construct P' as shown in Figure 13.2.

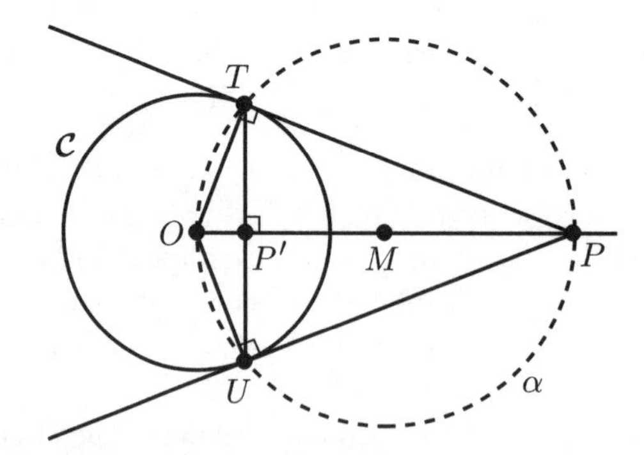

Figure 13.2. Contruction of P' if P is outside of $\mathcal{C}$

Exercises

*13.1.4. Carry out the construction in GeoGebra. Measure OP, OP', and OT and verify that $(OP)(OP') = (OT)^2$.

13.1.5. Prove that P' constructed above is the inverse of P.

*13.1.6. Use the construction to make a tool that finds $P' = I_{O,r}(P)$ for each point P outside $\mathcal{C}$.

The drawback to the tools produced in this section is that, while the first tool inverts points inside $\mathcal{C}$, a different tool is required to invert points that are outside $\mathcal{C}$. Fortunately GeoGebra provides one tool that inverts all points, whether they are inside or outside of $\mathcal{C}$. That tool will be introduced in the next section. The purpose of making the tools in this section is to deepen our understanding of inversion in a circle.

13.2 Inverting circles and lines

One reason inversions are useful is that they preserve certain features of Euclidean geometry. Inversions are not isometries, and they do not preserve such basic properties as distance and collinearity, but they do preserve other geometric relationships. The last section looked at what happens when a single point is inverted; this section explores what happens when all the points on a circle or line are inverted at once. We will see that inversions preserve circles and lines in the sense that the inverse of a circle is either a circle or a line and the inverse of a line is either a circle or a line.

We are accustomed to transformations that operate on one point at a time, but now we are applying a transformation to an entire set of points at once. To make certain that there is no misunderstanding, let us say exactly what we mean. For example, if $\mathcal{C} = \mathcal{C}(O, r)$ is a circle and α is a second circle, the *image of α under inversion in $\mathcal{C}$* (also called the inverse of α) is

$$I_{O,r}(\alpha) = \{P' \mid P' = I_{O,r}(P) \text{ for some } P \in \alpha\}.$$

A GeoGebra tool for inversion. In the GeoGebra tool bar there is a tool, Reflect Object about Circle, that performs inversion in a circle. It will find the image of any point or of an object such as a circle or line. It is simple to use: just select the tool, click on the object to invert, and then click on the circle of inversion.

Exercises

*13.2.1. Construct a circle $\mathcal{C}(O, r)$ and a second circle α such that $O \notin \alpha$. Use the Reflect Object about Circle tool to confirm that $I_{O,r}(\alpha)$ is a circle.

(a) Where is the circle $I_{O,r}(\alpha)$ if α is completely outside $\mathcal{C}$?

(b) Where is the circle $I_{O,r}(\alpha)$ if α intersects $\mathcal{C}$ in two points?

(c) Where is the circle $I_{O,r}(\alpha)$ if α is completely inside $\mathcal{C}$?

(d) For which circles α is $I_{O,r}(\alpha) = \alpha$?

*13.2.2. Construct a circle $\mathcal{C}(O, r)$ and a circle α such that $O \in \alpha$. Verify that $I_{O,r}(\alpha)$ is a line.

(a) How is the line $I_{O,r}(\alpha \setminus \{O\})$ related to $\mathcal{C}$ and α if $\mathcal{C}$ and α intersect in two points? (Draw a sketch.)

(b) How is the line $I_{O,r}(\alpha \setminus \{O\})$ related to $\mathcal{C}$ and α if $\mathcal{C}$ and α intersect in one point? (Draw a sketch.)

(c) How is the line $I_{O,r}(\alpha \smallsetminus \{O\})$ related to $\mathcal{C}$ and α if α is contained inside $\mathcal{C}$?.

(d) Construct a movable point P on α and its inverse P' on $I_{O,r}(\alpha)$. What happens to P' as P approaches O?

*13.2.3. Construct a circle $\mathcal{C}(O, r)$ and a line ℓ such that $O \notin \ell$. Confirm that $I_{O,r}(\ell)$ is a circle through O.

(a) How is the circle $I_{O,r}(\ell \cup \{\infty\})$ related to $\mathcal{C}$ if ℓ is outside $\mathcal{C}$?

(b) How is the circle $I_{O,r}(\ell \cup \{\infty\})$ related to $\mathcal{C}$ if $\mathcal{C}$ and ℓ intersect in two points?

(c) Construct a movable point P on ℓ and its inverse $P' \in I_{O,r}(\ell)$. What happens to P' as P approaches ∞?

(d) What is $I_{O,r}(\ell)$ if ℓ is a line that passes through O?

The following theorem summarizes the observations in the last three exercises.

Theorem. *Let $I_{O,r}$ be inversion in the circle $\mathcal{C} = \mathcal{C}(O, r)$.*
Part 1. *If α is a circle that does not pass through O, then $I_{O,r}(\alpha)$ is a circle that does not pass through O.*
Part 2. *If α is a circle that passes through O, then $I_{O,r}(\alpha \smallsetminus \{O\})$ is a line that does not pass through O.*
Part 3. *If ℓ is a line that does not pass through O, then $I_{O,r}(\ell \cup \{\infty\})$ is a circle that passes through O.*
Part 4. *If ℓ is a line that passes through O, then $I_{O,r}(\ell \cup \{\infty\}) = \ell \cup \{\infty\}$.*

Proof. See [11], Theorems 10.7.4 through 10.7.7. □

13.3 Othogonality

Orthogonality is another geometric relationship preserved by inversions.

Definition. Two circles are said to be *orthogonal* if they intersect and their tangent lines are perpendicular at the points of intersection.

Exercises

*13.3.1. Construct a circle $\mathcal{C} = \mathcal{C}(O, r)$ and a second circle α that is orthogonal to $\mathcal{C}$. Verify that $I_{O,r}(\alpha) = \alpha$. The two points of $\mathcal{C}(O, r) \cap \alpha$ determine two arcs on α. Verify that inversion in $\mathcal{C}(O, r)$ interchanges the two arcs.

*13.3.2. Construct a circle $\mathcal{C}(O, r)$, a point P not on $\mathcal{C}(O, r)$ and not equal to O, and its inverse $P' = I_{O,r}(P)$. Verify that if α is a circle that contains both P and P', then α is orthogonal to $\mathcal{C}(O, r)$.

Here is a statement of the theorem you verified in the last two exercises.

Theorem. *Let $\mathcal{C} = \mathcal{C}(O, r)$ and α be two circles.*
Part 1. *If α is orthogonal to $\mathcal{C}$, then $I_{O,r}(P) \in \alpha$ for every $P \in \alpha$.*
Part 2. *If there exists a point $P \in \alpha$ such that $I_{O,r}(P) \in \alpha$ and $I_{O,r}(P) \neq P$, then α is orthogonal to $\mathcal{C}$.*

Proof. See [11], Theorems 10.7.9 through 10.7.11. □

The theorem will be used in the exercises below to construct circles that are orthogonal to a fixed circle $\mathcal{C}$. The first construction allows us to specify two points that are to lie on the orthogonal circle. The second construction allows us to specify one point that lies on the circle and the tangent line to the circle at that point. Both tools will be used extensively in the next chapter.

Construction. Let $\mathcal{C} = \mathcal{C}(O, r)$ be a circle. Let A and B be two points such that O, A, B are noncollinear and A and B do not lie on $\mathcal{C}$. Construct $A' = I_{O,r}(A)$ and then construct α, the circumcircle for $\triangle ABA'$. The circle α contains A and B by construction; it is orthogonal to $\mathcal{C}$ by Part 2 of the theorem.

Exercise

*13.3.3. Assume $\mathcal{C} = \mathcal{C}(O, r)$ is a circle and that A and B are two points such that O, A, B are noncollinear and A and B are inside $\mathcal{C}$. Use the construction to create a GeoGebra tool that constructs a circle α that goes through the points A and B and is orthogonal to $\mathcal{C}$. The tool should accept four points as inputs (the center of inversion O, a point R on the circle of inversion, and the two points A and B), and return the circle α as output object.

What happens to α when you move the two points so that they lie on a diameter of $\mathcal{C}$?

Construction. Let $\mathcal{C} = \mathcal{C}(O, r)$ be a circle. Let t be a line and let $P \neq O$ be a point that lies on t but not $\mathcal{C}$. Construct the line ℓ that is perpendicular to t at P. Construct $P' = I_{O,r}(P)$ and $Q = \rho_\ell(P')$, the reflection of P' in ℓ. Assume O does not lie on ℓ so that Q is distinct from P'. Construct the circumcircle α of P, P', and Q (see Figure 13.3).

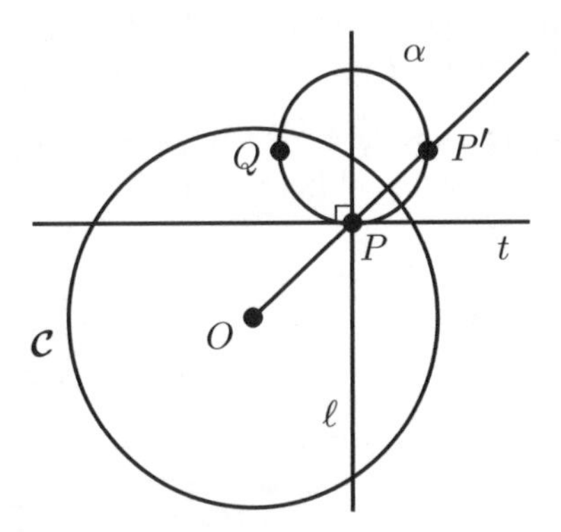

Figure 13.3. Construction of an orthogonal circle that is tangent to t at P

Exercises

13.3.4. Prove that the circle α in the construction is tangent to t and orthogonal to $\mathcal{C}$. Explain how to construct a circle with these properties if O lies on ℓ.

*13.3.5. Construct a circle $\mathcal{C} = \mathcal{C}(O, r)$, a point P not on $\mathcal{C}$, and a line t such that P lies on t. Assume $P \neq O$ and $\overleftrightarrow{OP}$ is not perpendicular to t. Create a GeoGebra tool that constructs a circle α such that α is tangent to t at P and is orthogonal to $\mathcal{C}$. The tool should accept four points as inputs (the center O of inversion, a point R on the circle of inversion, the point P and a second point on t), and return the circle α as output.

What happens to α as P approaches O?

13.4 Angles and distances

Given two intersecting lines in the plane, the inverses of the lines will be two intersecting circles. Inversions preserve angles in the sense that the angles between the lines are congruent to the angles between the tangent lines of the circles.

Inversions do not preserve individual distances, but they do preserve a combination of distances, the cross ratio.

Definition. Let A, B, P, and Q be four distinct points. The *cross-ratio* $[AB, PQ]$ of the four points is defined by

$$[AB, PQ] = \frac{(AP)(BQ)}{(AQ)(BP)}.$$

Exercises

*13.4.1. Construct two intersecting lines ℓ and m. Use the Reflect Object about Circle tool to invert the two lines. The result should be two circles that intersect at two points, one of which is O. Construct the two tangent lines at the other point of intersection. Measure the angles between ℓ and m and then measure the angles between the tangent lines to ℓ' and m'. Are the measures equal?

*13.4.2. Construct four points A, B, P, and Q. Measure the distances and calculate $[AB, PQ]$. Now invert the four points in a circle $\mathcal{C}(O, r)$ and calculate $[A'B', P'Q']$. Verify that $[AB, PQ] = [A'B', P'Q']$.

*13.4.3. Make a tool that calculates the cross ratio of four points.
Hint: A text box can be the output object for a tool.

14

The Poincaré Disk

In this final chapter we study the Poincaré disk model for hyperbolic geometry. Even though hyperbolic geometry is a non-Euclidean geometry, the topic is nonetheless appropriate for inclusion in a treatment of Euclidean geometry because the Poincaré disk is built within Euclidean geometry. The main tool used in the construction of the Poincaré disk model is inversion in Euclidean circles, so the tools you made in Chapter 13 will be used in this chapter to perform hyperbolic constructions. Many of the constructions in this chapter were inspired by those in the beautiful paper [7] by Chaim Goodman-Strauss.

It was Eugenio Beltrami (1835–1900) who originated the idea of representing hyperbolic geometry within Euclidean geometry. There are many ways in which to construct models of hyperbolic geometry, but the Poincaré disk model is probably the best known. One reason for its popularity is the great beauty of the diagrams that can be constructed in it. The model is named for the French mathematician Henri Poincaré (1854–1912) who first introduced it.

14.1 The Poincaré disk model for hyperbolic geometry

A *model* for a geometry is an interpretation of the technical terms of the geometry (such as point, line, distance, angle measure, etc.) that is consistent with the axioms of the geometry. The usual model for Euclidean geometry is $\mathbb{R}^2$, the Cartesian plane, which consists of all ordered pairs of real numbers. That model has been more or less assumed throughout this book, although most of the proofs we have given have been *synthetic*, which means that they are based on the axioms of geometry and not on specific (analytic) properties of any particular model for geometry.

We are about to begin the study of one form of non-Euclidean geometry—geometry in which the Euclidean parallel postulate is not assumed to hold. We will consider two kinds of non-Euclidean geometry: *neutral geometry* is the geometry that is based on all the usual axioms of Euclidean geometry except that no parallel postulate is assumed while *hyperbolic geometry* is the geometry that is based on all the axioms of neutral geometry together with the hyperbolic parallel postulate. The *hyperbolic parallel postulate* asserts that for every line ℓ and for every point P that does not lie on ℓ, there exist multiple lines through P that are parallel to ℓ. Since all the axioms of neutral geometry are also assumed in both Euclidean and hyperbolic geometries, any theorem that can be proved in neutral geometry is a theorem in both the other geometries as well. This is significant because many of the elementary constructions at the beginning of Euclid's *Elements* are neutral, so we can use them in our study of hyperbolic geometry.

Let us now describe the Poincaré disk model for hyperbolic geometry. Fix a circle Γ in the Euclidean plane. A "point" in the Poincaré disk is a Euclidean point that is inside Γ. There are two kinds of "lines" in the Poincaré disk. The first kind of line is a diameter of Γ; more specifically, a Poincaré line of the first kind consists of all the points on a diameter of Γ that lie inside Γ. A second kind of Poincaré line is a Euclidean circle that is orthogonal to Γ; more specifically, a Poincaré line of the second kind consists of all the points of a Euclidean circle orthogonal to Γ that lie inside Γ.

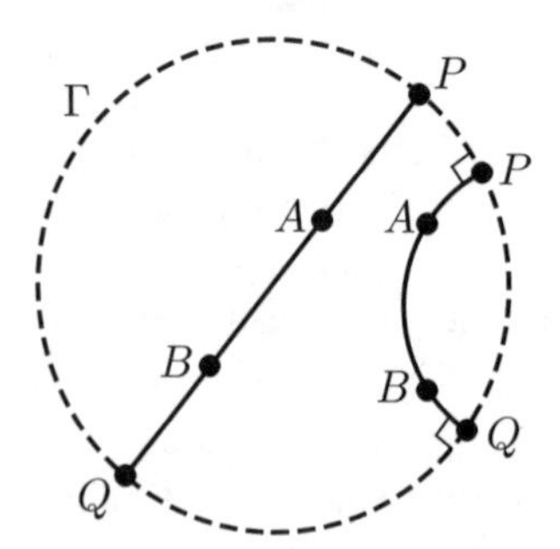

Figure 14.1. Two kinds of Poincaré lines and their ideal endpoints

The Poincaré distance between two points A and B in the Poincaré disk is defined by $d(A, B) = |\ln([AB, PQ])|$, where P and Q are the points at which the Poincaré line containing A and B intersects Γ. The points P and Q lie on Γ so they are not themselves points in the Poincaré disk, but they are useful in defining the distance between points in the Poincaré disk. The points on Γ are ideal points for the Poincaré disk. (In hyperbolic geometry, each line has two ideal endpoints and the set of ideal points forms a circle—see Figure 14.1.) The angle between two Poincaré lines is measured by measuring the Euclidean angle between their tangent lines.

The Poincaré disk is a model for hyperbolic geometry. Proving this assertion means proving that, with the terms point, line, distance, etc. interpreted as above, all the axioms of hyperbolic geometry are satisfied. Since the model is described within Euclidean geometry, those proofs are all Euclidean proofs. For example, we will see in the next section that theorems about Euclidean inversions in circles can be used to construct a unique Poincaré line through any two distinct points in the Poincaré disk. Thus Euclid's first postulate holds in the model. Also, Euclidean inversions preserve angles and cross-ratios (by the theorems in Chapter 13), so they preserve Poincaré angle measure and Poincaré distance. The Euclidean inversions are therefore isometries of the model and they function as reflections across lines in hyperbolic geometry. Inversions are the main tool used in proofs of theorems about the Poincaré disk.

In this chapter we will focus on constructions in the Poincaré disk and will not supply proofs that the axioms of hyperbolic geometry are satisfied in the model. Such proofs can be found in most college geometry texts. In particular, §13.3 of [11] contains a more detailed description of the model and proofs that all the usual axioms of geometry are satisfied by this model.

14.2 The hyperbolic straightedge

Our first objective is to make a hyperbolic straightedge. That is, we want to make a GeoGebra tool that constructs the unique Poincaré line through two points in the Poincaré disk. We also want to make variations on the tool that will construct the Poincaré ray or segment determined by the points. The tool you made in Exercise 13.3.3 is almost exactly what is needed in order to construct a Poincaré line; the only refinement required is to trim off the part of the Euclidean circle that lies outside Γ.

When you make the tools in this and later sections, you may assume that the two given points do not lie on a common diameter of Γ. This means that the tools you make will only construct Poincaré lines of the second kind. As a result, some of the lines may disappear momentarily when you move one point across the diameter determined by another. Making the tools general enough that they can accept any two points as inputs would significantly increase the complication of the constructions, so we will not bother with it. One justification for this is that randomly chosen points would not lie on a common diameter, so the case we are omitting is the rare, exceptional case. Another justification is that the simple tools produce a good approximation to the correct answer anyway since part of a circle of large radius is indistinguishable from a straight line.

Exercises

*14.2.1. Make a tool that constructs the Poincaré line through two points A and B in the Poincaré disk. The tool should accept four points as inputs (the center O of Γ, a point R on Γ, and the two points A and B) and return the hyperbolic line ℓ as its output. Assume in your construction that A, B, and O are noncollinear. What happens to ℓ when you move A to make the three points collinear?
Hints: Remember that the Poincaré "line" ℓ is an arc of a Euclidean circle α. First use the tool you created in Exercise 13.3.3 to construct the Euclidean circle α that contains A and B and is orthogonal to Γ. Mark the two points of $\alpha \cap \Gamma$. Then use the Circumcircular Arc through Three Points tool to define just the arc of α that lies inside Γ.

*14.2.2. Use the tool from the last exercise to construct a Poincaré line ℓ. Construct a point P that does not lie on ℓ. Construct multiple lines through P that are all parallel to ℓ. Is there any limit to the number of parallel lines you can construct?
Hint: Remember that "parallel" simply means that the lines do not intersect.

*14.2.3. Make a tool that constructs the Poincaré segment determined by two points A and B. The tool should accept four points as inputs (the center O of Γ, a point R on Γ, and the two points A and B) and return the hyperbolic segment from A to B as its output. The tool you construct in this exercise will be referred to as the *hyperbolic straightedge*.
Hint: The Poincaré segment is again an arc of the Euclidean circle α that contains A and B and is orthogonal to Γ. It is tricky to get GeoGebra to consistently choose

the correct arc of this circle. One way that seems to work is to begin by constructing α; then construct the Euclidean segment $\overline{AB}$, find its midpoint M, construct the Euclidean segment $\overline{OM}$, mark the point C at which $\overline{OM}$ crosses α (which is necessarily between A and B), and define the Poincaré segment from A to B to be the Euclidean circular arc determined by the three points A, C, and B.

*14.2.4. Use your hyperbolic straightedge to construct a triangle in the Poincaré disk. Drag the vertices around to see what shapes are possible.

*14.2.5. Make a tool that constructs the Poincaré ray determined by two points A and B in the Poincaré disk. The tool should accept four points as inputs (the center O of Γ, a point R on Γ, and the two points A and B) and return the hyperbolic ray from A through B as its output.
Hint: Once again the Poincaré ray is an arc of the Euclidean circle α that contains A and B and is orthogonal to Γ. The difficult task is to give GeoGebra instructions that will allow it to consistently choose the correct arc of α. I have not found a way to do this that reliably passes the drag test. The following construction comes close. Construct the inverse B' of B in Γ, and construct the circular arc determined by A, B, and B'. Then construct the point Q where the arc crosses Γ. The circular arc determined by A, B, and Q is the Poincaré ray we want. The construction consistently produces the correct ray. Unfortunately it does not hold up perfectly when subjected to the drag test. The problem is that the ray disappears when one of the points A and B is moved across the diameter of Γ determined by the other point. I have not found a way to overcome this problem. An acceptable but less than perfect solution to the problem is to use the circular arc determined by the points A, B, and B'. This is unsatisfying because it gives a longer arc than we want. But the extra piece is outside the Poincaré disk and does not cause any harm to the portion of the sketch that is inside Γ, and that is the part of the diagram in which we are most interested.

14.3 Common perpendiculars

In the hyperbolic plane there are two kinds of parallel lines [11, §8.4]. One kind occurs when two parallel lines m and n are *asymptotically parallel*, which means that in one direction the two lines get closer and closer together. This is illustrated in the Poincaré disk model by Poincaré lines that approach a common (ideal) point of Γ. Figure 14.2 shows three Poincaré lines that are pairwise asymptotically parallel. Even though the lines

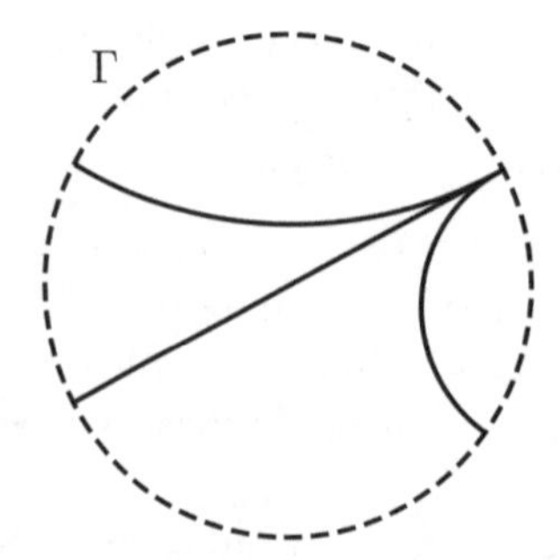

Figure 14.2. Three asymptotically parallel lines

converge at Γ, they are parallel (disjoint) because they do not intersect at a point of the Poincaré disk.

The classification of parallels theorem [11, Theorem 8.4.18] asserts that if m and n are two lines that are parallel but not asymptotically parallel, then m and n must admit a common perpendicular. This means that there is a line t that is perpendicular to both m and n (see Figure 14.3). In the next set of exercises you will construct the common perpendicular line. The first five exercises are exercises in Euclidean geometry. The sixth exercise applies the Euclidean constructions to the Poincaré disk. The constructions in this section are based on the ideas in [10].

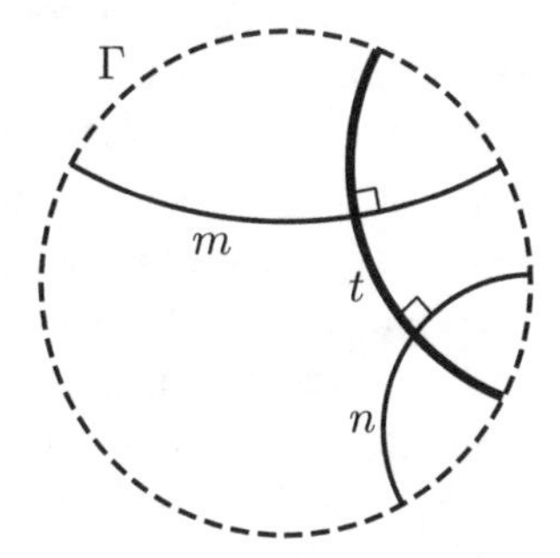

Figure 14.3. The Poincaré line t is the common perpendicular for m and n

Exercises

14.3.1. Let α be a circle with center A and let C be a point outside α. Construct the circle δ with diameter $\overline{AC}$ and let H be a point of $\alpha \cap \delta$. Define γ to be the circle with center C and radius $\overline{CH}$. Use Thales' theorem to prove that α and γ are orthogonal.

*14.3.2. Use the construction of Exercise 14.3.1 to make a tool that constructs the circle that is orthogonal to a specified circle α and has a specified point C as center. The tool should accept three points as inputs (the center A of α, a point S on α, and a point C outside α) and should return a circle γ that has C as its center and is orthogonal to α as its output.

14.3.3. Let α and β be two orthogonal circles and let ℓ be their common secant line. Let γ be a circle whose center lies on ℓ. Use the theorem in §13.3 to prove that γ is orthogonal to α if and only if γ is orthogonal to β.

14.3.4. Let α, β, and Γ be three circles such that α and β are disjoint and each of α and β is orthogonal to Γ. Let m be the common secant line for α and Γ, and let n be the common secant line for β and Γ. Assume $m \parallel n$. Prove that α and β are orthogonal to a diameter of Γ.

14.3.5. Let α, β, and Γ be three circles such that α and β are disjoint and orthogonal to Γ. Let m be the common secant line for α and Γ, let n be the common secant line for β and Γ, and let P be the point at which m and n intersect. By Exercise 14.3.1 there is a circle τ with center P that is orthogonal to Γ. Use Exercise 14.3.3 to prove that τ is also orthogonal to both α and β.

*14.3.6. Construct two Poincaré lines ℓ and m that are parallel but not asymptotically parallel. Use Exercise 14.3.5 to construct a Poincaré line t such that t is orthogonal to both ℓ and m. Under what conditions will the common perpendicular be a Poincaré line of the first kind (a diameter of Γ)?

14.4 The hyperbolic compass

Our next task is to make a GeoGebra tool that will function as a hyperbolic compass. (In this context "compass" is the instrument traditionally used in the compass and straightedge constructions of elementary geometry, not the kind of compass we use to tell direction when we are out hiking.) The tool should construct the Poincaré circle with a specified point A as center and and a specified point B on the circle. The Poincaré circle with center A and radius r is defined to be the set of all points X in the Poincaré disk such that $d(A, X) = r$. So the obvious way to attempt to construct a Poincaré circle is to make a tool that measures Poincaré distances and then construct the circle as the locus of all points at a fixed Poincaré distance from the center. This method is not easy to implement and there is a more elegant way to construct the hyperbolic circle.

The construction is based on two facts. The first is that every Poincaré circle is also a Euclidean circle [11, Exercises 13.7–13.9]. In general the Poincaré center of the circle is different from the Euclidean center and the Poincaré radius is different from the Euclidean radius, but for each Poincaré circle there is a point A' and a number r' such that the Poincaré circle is exactly equal to the Euclidean circle $\mathcal{C}(A', r')$. Thus we need only locate the Euclidean center A' and then construct the Euclidean circle with center A' that passes through B.

The second fact the construction is based on is that the tangent line theorem (Chapter 0) is a theorem in neutral geometry. As a result, the theorem holds in the Poincaré disk and a Poincaré circle α must be perpendicular to every Poincaré line through the Poincaré center A. In particular, α must be perpendicular to the Euclidean line $\overleftrightarrow{OA}$, where A is the Poincaré center of the circle and O is the center of the Euclidean circle Γ that is used in the definition of the Poincaré disk (since it is a Poincaré line of the first kind). This means that the Euclidean center must lie on $\overleftrightarrow{OA}$. In addition, α must be perpendicular to the Poincaré line determined by A and B. This implies that the Euclidean center must lie on line that is tangent to the Poincaré line at B.

Construction. Assume that the Poincaré disk is defined by the Euclidean circle Γ with center O. Let points A and B inside Γ be given. Assume that A, B, and O are noncollinear. Construct the Euclidean line $\overleftrightarrow{OA}$. Construct the Poincaré line ℓ that contains A and B. Let t be the Euclidean line that is tangent to ℓ at B. Define A' to be the point at which t and $\overleftrightarrow{OA}$ intersect. (See Figure 14.4.) The Euclidean circle with center at A' and passing through B is the Poincaré circle with center A passing through B.

Exercises

*14.4.1. Make a tool that constructs the Poincaré circle determined by two points in the Poincaré disk. The tool should accept four points as inputs (the center O of Γ, a

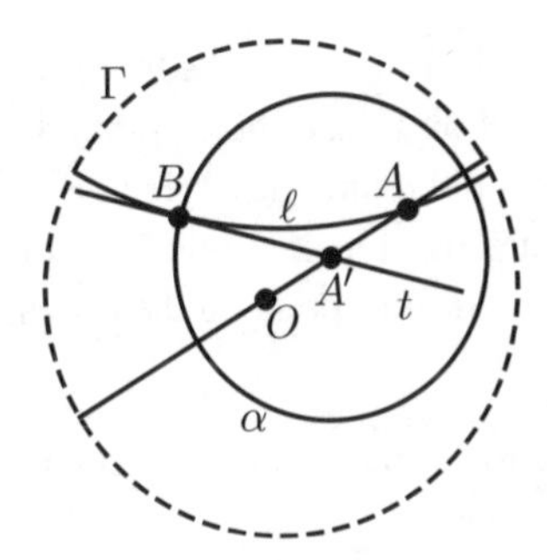

Figure 14.4. Construction of the Poincaré circle through B with center A

point R on Γ, a point A that is to be the center of the Poincaré circle, and a point B that is to lie on the Poincaré circle) and return the hyperbolic circle α as its output. Assume in your construction that A, B, and O are noncollinear. What happens to α when you move A to make the three points collinear? What happens when you move the "center" of the circle towards the boundary of the Poincaré disk? This tool you construct in this exercise will be referred to as the *hyperbolic compass*.

*14.4.2. Use your hyperbolic compass to construct a hyperbolic circle and then use your hyperbolic straightedge to construct a triangle inscribed in the circle. Construct an inscribed triangle that has a diameter as one side. Based on your diagram, do you think Thales' theorem is correct in hyperbolic geometry?

14.5 Other hyperbolic tools

Most of the compass and straightedge constructions you learned in high school are part of neutral geometry and are therefore valid in hyperbolic geometry. Since we now have a hyperbolic compass and straightedge, we can employ them with the high school constructions to make other hyperbolic tools.

Exercises

*14.5.1. Make a Poincaré perpendicular bisector tool. The tool should accept four points as inputs (the center O of Γ, a point R on Γ, and the two endpoints A and B of the Poincaré segment) and it should return a Poincaré line as its output.
Hint: The Pointwise Characterization of Perpendicular Bisectors (Chapter 0) is a neutral theorem. Thus you need only find two points that are equidistant from A and B and then take the Poincaré line through them. We have not yet measured Poincaré distances, but we do know how to construct Poincaré circles.

*14.5.2. Make a Poincaré midpoint tool.

*14.5.3. Make a Poincaré perpendicular line tool; i.e., a tool that drops a perpendicular from a point P to a Poincaré line ℓ. The tool should accept five points as inputs (the center O of Γ, a point R on Γ, two points A and B that determine a hyperbolic line ℓ, and a point P that is to lie on the perpendicular) and return a Poincaré line as output.
Hint: Use your Poincaré compass to find two points on ℓ that are equidistant from P. The perpendicular line is the perpendicular bisector of the segment connecting the two points.

*14.5.4. Make a Poincaré angle bisector tool. The tool should accept five points as inputs (the center O of Γ, a point R on Γ, and three points A, B, and C that define a hyperbolic angle $\angle ABC$) and should return the angle bisector, which is a hyperbolic ray. Hint: The pointwise Characterization of Angle Bisectors (Chapter 0) is a neutral theorem. You should be able to locate a point on the ray as the intersection of two hyperbolic circles.

*14.5.5. Use your hyperbolic tools to determine whether or not the secant line theorem is correct in hyperbolic geometry.

14.6 Triangle centers in hyperbolic geometry

We are now ready to explore triangle centers in hyperbolic geometry. Some of the classical triangle centers exist in hyperbolic geometry and others do not.

Ideally you will use the tools created earlier in the chapter to investigate the hyperbolic triangle centers in a way that is analogous to what you did earlier when you investigated Euclidean triangle centers. For example, you can use your hyperbolic straightedge to construct a hyperbolic triangle and use the hyperbolic angle bisector to locate its incenter. Then you can use the hyperbolic perpendicular line tool to drop a perpendicular from the incenter to one side of the triangle and thus locate a point on the incircle. In practice, however, there seem to be mysterious quirks in the software that prevent the hyperbolic tools from working consistently and reliably in this setting. If you find that to be so, you will have to do *ad hoc* constructions of the hyperbolic objects you need as you need them. You can view the work you did earlier in creating the tools to be practice in mastering the constructions involved.

Exercises

*14.6.1. The hyperbolic incenter and incircle.

(a) Construct a Poincaré triangle and the three bisectors of the interior angles of the triangle. Verify that the angle bisectors are always concurrent, regardless of the shape of the triangle.

(b) Construct the hyperbolic incircle for your triangle.

*14.6.2. Hyperbolic median concurrence.

(a) Construct a Poincaré triangle, the three midpoints of the sides, and the medians. Verify that the medians are concurrent regardless of the shape of the triangle. The point of concurrence is called the hyperbolic centroid of the triangle.

(b) Construct the hyperbolic medial triangle.

*14.6.3. The hyperbolic circumcenter and circumcircle.

(a) Construct a Poincaré triangle and the hyperbolic perpendicular bisectors of the three sides. Move the vertices of the triangle to vary the shape of the triangle. Verify that the perpendicular bisectors are concurrent for some triangles and not for others.

(b) Can you find a triangle for which exactly two of the perpendicular bisectors intersect? If so, describe it.

(c) Can you find a triangle for which all three of the perpendiculars are asymptotically parallel? If so, describe it.

(d) Can you find a triangle for which all three of the perpendiculars admit a common perpendicular? If so, describe it.

(e) Can you find a triangle for which exactly two of the perpendicular bisectors are asymptotically parallel? If so, describe it.

(f) Verify that the triangle has a hyperbolic circumcircle if the three perpendicular bisectors are concurrent.

14.6.4. Prove that if two of the hyperbolic perpendicular bisectors of the sides of a hyperbolic triangle intersect, then the third perpendicular bisector also passes through the point of intersection.

*14.6.5. The hyperbolic orthocenter.

(a) Construct a Poincaré triangle and the three hyperbolic altitudes. Verify that the altitudes are concurrent for some triangles and not for others.

(b) Can you find a triangle for which exactly two of the altitudes intersect? If so, describe it.

(c) Can you find a triangle for which all three of the altitudes are asymptotically parallel? If so, describe it.

(d) Can you find a triangle for which all three of the altitudes admit a common perpendicular? If so, describe it.

(e) Can you find a triangle for which exactly two of the altitudes are asymptotically parallel? If so, describe it.

(f) Construct the hyperbolic orthic triangle.

*14.6.6. The hyperbolic Euler line.

Construct a Poincaré triangle that has both a hyperbolic circumcenter and a hyperbolic orthocenter. Let us agree to call the Poincaré line determined by the circumcenter and the orthocenter the *hyperbolic Euler line* for the triangle (provided the circumcenter and orthocenter exist). Now construct the hyperbolic centroid of the triangle. Does the hyperbolic centroid always lie on the hyperbolic Euler line?

References

[1] Claudi Alsina and Roger B. Nelsen. *Icons of Mathematics*. The Dolciani Mathematical Expositions, volume 45. The Mathematical Association of America, Washington, DC, 2011.

[2] Conference Board of the Mathematical Sciences CBMS. *The Mathematical Education of Teachers*. Issues in Mathematics Education, volume 11. The Mathematical Association of America and the American Mathematical Society, Washington, DC, 2001.

[3] H. S. M. Coxeter and S. L. Greitzer. *Geometry Revisited*. New Mathematical Library, volume 19. The Mathematical Association of America, Washington, DC, 1967.

[4] Dana Densmore, editor. *The Bones: A Handy, Where-to-find-it Pocket Reference Companion to Euclid's Elements*. Green Lion Press, Santa Fe, New Mexico, 2002.

[5] Dana Densmore, editor. *Euclid's Elements*. Green Lion Press, Santa Fe, New Mexico, 2002.

[6] William Dunham. *Euler: The Master of Us All*. The Dolciani Mathematical Expositions, volume 22. The Mathematical Association of America, Washington, DC, 1999.

[7] Chaim Goodman-Strauss. Compass and straightedge in the Poincaré disk. *American Mathematical Monthly*, 108:38–49, 2001.

[8] Sir Thomas L. Heath. *The Thirteen Books of Euclid's Elements with Introduction and Commentary*. Dover Publications, Inc., Mineola, New York, 1956.

[9] I. Martin Isaacs. *Geometry for College Students*. The Brooks/Cole Series in Advanced Mathematics. Brooks/Cole, Pacific Grove, CA, 2001.

[10] Michael McDaniel. The polygons are all right. *preprint*, 2006.

[11] Gerard A. Venema. *The Foundations of Geometry*. Pearson Education, Inc., Boston, second edition, 2012.

Index

*-d exercises, viii, 13

acute angle, 2
additivity of area, 12
advanced Euclidean geometry, vii
alternate interior angles theorem, 6
altitude, 25
    concurrence, 72
altitude concurrence theorem, 26
angle, 2
angle addition postulate, 2
angle bisector, 4
    concurrence, 73
angle bisector concurrence theorem, 41
angle sum theorem, 7
angle-angle-side theorem, 4
angle-side-angle theorem, 3
anticomplementary triangle, 48
antimedial, 48
arc intercepted by, 10
Archimedes, 44
area, 11
    in GeoGebra, 19

Beltrami, E., 111
betweenness
    for points, 1
    for rays, 2
bisect, 20
Brahmagupta's formula, 56
Brianchon's theorem, 90, 91
Brianchon, C., 57, 78, 90
bride's chair, 35
butterfly theorem, 97

Cabri Geometry, viii
calculation
    in GeoGebra, 18
Cartesian plane, 111
center
    of a circle, 10
    of a square, 100
    of mass, 25
central angle, 10
central angle theorem, 10
centroid, 24, 72
    hyperbolic, 118
Ceva's theorem
    standard form, 51, 68
    trigonometric form, 71
Ceva, G., 50, 63
Cevian
    line, 50, 63
        proper, 63
    triangle, 50
check boxes, 33
Cinderella, viii
circle, 10
circle-circle continuity, 5
circle-line continuity, 5
circumcenter, 27, 39, 72
    hyperbolic, 118
circumcircle, 39
    hyperbolic, 118
circumradius, 39
circumscribed
    about a circle, 90
    circle, 39
classical triangle center, 23
coaxial, 86
concave quadrilateral, 54
concurrence theorems, 72
concurrency problem, 63

concurrent, 23
congruent
  angles, 2
  quadrilaterals, 9, 53
  segments, 1
  triangles, 3
conjugate, 73
convex quadrilateral, 54
copolar, 86
corresponding angles, 6
corresponding angles theorem, 7
corresponding central angle, 10
create a tool, 31
cross-ratio, 110
crossbar theorem, 5, 69
crossed quadrilateral, 54
cyclic quadrilateral, 55

de Longchamps point, 48
dependent objects, 15
Desargues's theorem, 86
Desargues, G., 77, 86
diagonal
  of a quadrilateral, 9, 56
diameter
  of a circle, 10
directed
  angle measure, 71
  distance, 66
distance, 1
drag test, 17
drop a perpendicular, 6
duality, 77

elementary Euclidean geometry, 1
equicircle, 42
equilateral triangle, 16
escribed circle, 42
Euclid, vii
Euclid's *Elements*, vii, 1
Euclid's fifth postulate, 6
Euclid's Proposition
  I.1, 16
  I.12, 6
  I.15, 3
  I.16, 5
  I.26, 3
  I.27, 6
  I.28, 7
  I.29, 6
  I.32, 7
  I.34, 9
  I.4, 3
  I.47, 35
  I.5, 4
  I.8, 3
  III.20, 10
  III.21, 10
  III.22, 55
    converse, 56
  III.3, 11
  III.31, 11
    converse, 11
  III.32, 81
  III.36, 81
  VI.2, 8
  VI.31, 35
  VI.4, 8
  VI.6, 8
Euclidean geometry, vii
Eudoxus, 8
Euler line, 28, 60
  hyperbolic, 119
Euler line theorem, 28
Euler, L., 23, 27, 57
excenter, 42, 73
excircle, 42
extended
  Euclidean plane, 66
extended law of sines, 40, 52
exterior angle, 5
exterior angle theorem, 5
external tangents theorem, 11

Fermat point, 101
Fermat, P., 101
Feuerbach
  point, 60
  triangle, 60
Feuerbach's theorem, 60
Feuerbach, K. W., 57
foot
  of a perpendicular, 6
  of an altitude, 25
free objects, 15

general position, 96
GeoGebra, viii, 13
  area, 19
  calculation, 18
  check boxes, 33
  create a tool, 31
  measuring tools, 18
  save a tool, 32
  toolbar, 13
  user-defined tools, 31

Geometer's Sketchpad, viii
Gergonne point, 43, 73
Gergonne, J., 44

Heron, 44
hexagon, 88
hyperbolic
  compass, 116
  geometry, 111
  parallel postulate, 112
  straightedge, 113
hypotenuse-leg theorem, 4

ideal point, 66
image of a set, 107
incenter, 42, 73
  hyperbolic, 118
incident, 77
incircle, 41, 42
  hyperbolic, 118
inradius, 42
inscribed
  angle, 10
  circle, 41, 42
  hexagon, 88
  quadrilateral, 55
inscribed angle theorem, 10, 56
interior
  angle, 5
  of an angle, 2
internal angle bisector, 29
inversion in a circle, 105
inversive plane, 105
involution, 73
isogonal
  conjugate, 74
  of a Cevian line, 74
isosceles triangle theorem, 4
isotomic conjugate, 73

Jordan curve theorem, 54

Lehmus, C., 29
length of a segment, 1
line, 1
  at infinity, 66
Line tools, 14
  more, 16
linear pair, 2
linear pair theorem, 2

mark a point, 14
measure of an angle, 2
measuring tools, 18
medial triangle, 47
  hyperbolic, 118
median, 24
  concurrence, 72
  hyperbolic, 118
median concurrence theorem, 24
Menelaus of Alexandria, 77
Menelaus point, 78
  proper, 78
Menelaus's theorem
  standard form, 51, 79
  trigonometric form, 80
midpoint, 7
  quadrilateral, 20, 55
Miquel point, 101
Miquel's theorem, 101
Miquel, A., 101
model for a geometry, 111
Morley triangle, 102
Morley's theorem, 103
Morley, F., 103
movable point, 15
Move tool, 14

Nagel point, 44, 73
Nagel, C., 44
Napoleon
  point, 100
  triangle, 100
Napoleon Bonaparte, 99
Napoleon's theorem, 100
neutral geometry, 112
New Point tool, 14
nine-point
  center, 59
  circle, 57
nine-point center theorem, 60
nine-point circle theorem, 57
non-Euclidean geometry, vii, 111

obtuse angle, 2
opposite sides of a quadrilateral, 9, 53
Options menu, 13
ordinary point, 66
orthic triangle, 48
  hyperbolic, 119
orthocenter, 26, 72
  hyperbolic, 119
orthogonal circles, 108

Pappus of Alexandria, 91
Pappus's theorem, 92
parallel, 6

parallel postulate, 6, 111
parallel projection theorem, 8
parallelogram, 9, 54, 56
Pascal line, 88
Pascal's mystic hexagram, 88
Pascal, B., 78, 88
Pasch's axiom, 4
Pasch, M., 4
pedal
- line, 94
- triangle, 51

perpendicular, 6
perpendicular bisector, 7
- concurrence, 72

perspective
- from a line, 86
- from a point, 86

perspector, 86
perspectrix, 86
Playfair's postulate, 6
Poincaré disk model, 111
Poincaré, H., 111
point
- at infinity, 66, 105
- of concurrency, 23
- of perspective, 86

Point tools
- other, 14

pointwise characterization
- of angle bisector, 4
- of perpendicular bisector, 7

pole
- of a Simson line, 94

polygon, 11
Polygon tools, 16
Poncelet, J., 57
power of a point, 81, 82
projective geometry, 77
proper
- Cevian line, 63
- Menelaus point, 78

prove, ix
Ptolemy, 96
Ptolemy's theorem, 96
Pythagoras, 7
Pythagorean theorem, 8, 34

quadrangle, 53
quadrilateral, 9, 53

radical
- axis, 83
- center, 84

radical axis theorem, 84
radical center theorem, 84
radius
- of a circle, 10

ray, 1
rectangle, 9, 56
region, 11
rhombus, 9, 56
right angle, 2

SAS similarity criterion, 8
save a tool, 32
secant line theorem, 11
second pedal triangle, 51
segment, 1
semiperimeter, 44
sensed ratio, 66
side
- of a polygon, 11
- of a quadrilateral, 9, 53
- of a triangle, 3, 25

side-angle-side theorem, 3
side-side-side theorem, 4
sideline, 25
similar triangles, 8
similar triangles theorem, 8
Simson line, 52, 94
Simson's theorem, 52, 93
Simson, R., 93
square, 9, 56
standard form
- of Ceva's theorem, 68
- of Menelaus's theorem, 79

Steiner, J., 29
Steiner-Lehmus theorem, 29
supplements, 2
symmedian, 76
- point, 76

synthetic, 111

tangent line, 10, 85
tangent line theorem, 10
Thales' theorem, 11
third pedal triangle, 51
toolbar, 13
topology, 54
Torricelli point, 100
Torricelli, E., 100, 101
transformation, 105
transversal, 6
trapezoid, 9
triangle, 3
- center, 23

hyperbolic, 118
congruence conditions, 3
triangle congruence condition
AAS, 4
ASA, 3
HL, 4
SAS, 3
SSS, 4
trigonometric form
of Ceva's theorem, 71
of Menelaus's theorem, 80
trilateral, 53, 78
trisect, 102
tritangent circle, 42

user-defined tools, 31

van Aubel's theorem, 100
Varignon's theorem, 55
Varignon, P., 55
Vecten, 35
Vecten configuration, 35
Vecten point, 36, 73
verify, ix
vertex
of a polygon, 11
of a quadrilateral, 9, 53
of a triangle, 3
vertical
angles, 2
pair, 2
vertical angles theorem, 3
View menu, 13

Wallace, W., 93

About the Author

Gerard Venema earned an A.B. in mathematics from Calvin College and a Ph.D. from the University of Utah. After completing his education he spent two years in a postdoctoral position at the University of Texas at Austin and another two years as a Member of the Institute for Advanced Study in Princeton, NJ. He then returned to his alma mater, Calvin College, and has been a faculty member there ever since. While on the Calvin College faculty he has also held visiting faculty positions at the University of Tennessee, the University of Michigan, and Michigan State University. He spent two years as Program Director for Topology, Geometry, and Foundations in the Division of Mathematical Sciences at the National Science Foundation.

Venema is a member of the American Mathematical Society and the Mathematical Association of America. He served for ten years as an Associate Editor of the *American Mathematical Monthly* and currently sits on the editorial board of MAA *FOCUS*. Venema has served the Michigan Section of the MAA as chair and is the 2013 recipient of the section's distinguished service award. He currently holds the position of MAA Associate Secretary and is a member of the Association's Board of Governors.

Venema is the author of two other books. One is an undergraduate textbook, *Foundations of Geometry*, published by Pearson Education, Inc., which is now in its second edition. The other is a research monograph, *Embeddings in Manifolds*, coauthored by Robert J. Daverman, that was published by the American Mathematical Society as volume 106 in its Graduate Studies in Mathematics series. In addition to the books, Venema is author of over thirty research articles in geometric topology.